*Sponsored by the Hebrew University
of Jerusalem and Harvard University Press*

THE JERUSALEM-HARVARD LECTURES

Freeman Dyson

IMAGINED
WORLDS

HARVARD UNIVERSITY PRESS
Cambridge, Massachusetts   London, England   1997

081
D

Library of Congress Cataloging-in-Publication Data
Dyson, Freeman J.
Imagined worlds / Freeman Dyson.
p. cm. — (Jerusalem-Harvard lectures)
Includes bibliographical references and index.
ISBN 0-674-53908-7
I. Title. II. Series.
AC8.D97 1997
081—dc20   96-31042

acknowledgments

This book grew out of a set of lectures given in May 1995 at the Hebrew University of Jerusalem, sponsored jointly by the Hebrew University and Harvard University Press. I am grateful to both institutions for their generous hospitality and support. I thank especially my hosts Hanoch Gutfreund, then the Rector of the Hebrew University, and Dorothy Harman, representative of Harvard University Press in Jerusalem. I am indebted to Michael Fisher, Susan Wallace Boehmer, and Ann Downer-Hazell, my editors at Harvard University Press, for their help and encouragement in converting the lectures into a book.

Fifteen years ago my friend June Goodfield published *An Imagined World,* a true story of a remarkable woman who combined the vocations of medical research and poetry. June Goodfield's book and mine have nothing in common except the titles. Our titles fit our themes, since her theme is singular while mine is plural.

October 1996

contents

INTRODUCTION

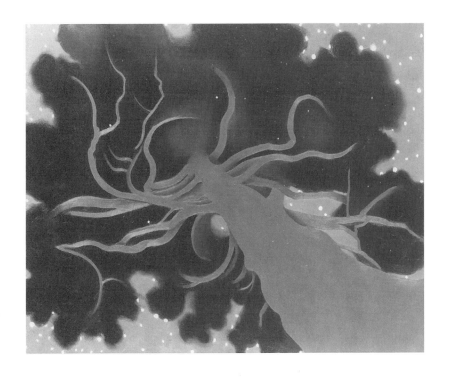

THE LAWRENCE TREE

By Georgia O'Keeffe, 1929. Reproduced by permission of Wadsworth Atheneum, Hartford. The Ella Gallup Sumner and Mary Catlin Sumner Collection Fund. Also by permission of the Artists Rights Society (ARS) New York.

ONKEL BRUNO WAS MY WIFE'S UNCLE, a country doctor who lived in a big house in a village in Germany. He inherited the house, along with the medical practice, from his father, and he stayed in it all his life. During that time, Germany was ruled by potentates of many stripes, imperial, Republican, National Socialist, and communist. Like the vicar of Bray, Onkel Bruno made his peace with whichever party was in power and carried on with his profession. I visited him at his home toward the end of his life, when he was a citizen of the German Democratic Republic. He expressed no enthusiasm for the communist society in which he lived but was grateful to the communists for leaving him in peace.

His magnificent house and garden were the pride and joy of his declining years. When I admired the large oak tree that stood in front of the house, Onkel Bruno said in a matter-of-fact tone, "That tree will have to come down; it has passed its prime." So far as I could see, the tree was in good health and showed no signs of imminent collapse. I asked him how he could dare to chop it down. He replied, "For the sake of the grand-children. That tree would last my time, but it would not last theirs. I will plant a tree that they will enjoy when they are as old as I am now." He expected his grandchildren to inherit his practice and live their lives in his home. That is the way it was in the world that he knew. Governments come and go but the family endures. You live for your children and for your grand-children. Horizons are long, and it is normal and natural to look ahead a hundred years, the time an oak tree takes to grow.

When I was a student in Cambridge, England, my college made a similar decision. The driveway to Trinity on the river side came through a magnificent avenue of elms planted in the eighteenth century. The elms were still beautiful but past their prime. The college decided, like Onkel Bruno, to sacrifice the present for the sake of the future. The avenue was chopped down and replaced by two rows of scrawny saplings. Now, fifty years later, the saplings are growing toward maturity. The avenue is again beautiful, and it will grow to full height as the twenty-first century goes by. Trinity College has been a great center of learning since it was founded in the sixteenth century, and it intends to remain a great center of learning in the twenty-first.

In October 1995 I attended a meeting in Slovenia, the East–West High-Tech Forum organized by my daughter. The purpose of the meeting was to allow leaders of the computer and software industries from East and West to meet and exchange ideas. Many people came from Russia and Eastern Europe, an equal number from America and Western Europe. All of them were doing well and expecting to do better. They were driving in the fast lane. The Easterners represented the new wave of business executives rising from the ashes of the old communist societies; the Westerners represented forward-looking businesses moving into the newly opened Eastern markets. The two sides shared certain basic assumptions: they believed that they were riding the wave of history; they believed that the triumph of free-market economics was inevitable and that they were helping to bring it about; and their horizons were short.

In the information world to which they belong, five years is a long time; fortunes are won and lost in a year or two. It makes no sense to make plans beyond five years, because the growth of information technology is unpredictable, and the workings of the free market are even more unpredictable. These new young capitalists grew up in a world of long-range socialist plans that failed, and they see no virtue in long-range plans of any kind. In all the discussions that I heard, the twenty-first century was hardly mentioned.

It seems that the modern world has grown increasingly short-sighted in recent years, as if the collapse of socialist economies and the victories of the free market have made all long-range visions of the future illusory. The voices of Onkel Bruno and Trinity College, striving to preserve small islands of natural beauty for our grandchildren, seem to be voices from the past, hardly audible amidst the intensifying winds of change. The public dialogue of our era is mainly a debate between free-market economists and conservationists, conservationists trying to preserve the past, free-market economists devaluing the future at a discount rate of seven percent per year. Neither side of the debate speaks for the future.

Who in the modern age still has dreams that extend beyond the lifetimes of our grandchildren? Two voices speak for the future, the voice of science and the voice of religion. Science and religion are two great human enterprises that endure through the centuries and link us with our descendants. I am a scientist, and as I attempt to look into the future in this book, I speak with the voice of science. I describe the past and the future from the scientific viewpoint that is familiar to me. But I do not claim that the voice of science speaks with unique authority. Religion has at least an equal claim to authority in defining human destiny. Religion lies closer to the heart of human nature and has a wider currency than science. Like the human nature that it reflects, religion is often cruel and perverted. When science achieved power to equal the power of religion, science often became cruel and perverted, too.

The poet W. H. Auden, who was a Christian, wrote of the importance of Christianity to the birth of modern literature in late antiquity: "One may like or dislike Christianity, but no one can deny that it was Christianity and the Bible which raised Western literature from the dead. A faith which held that the Son of God was born in a manger, associated himself with persons of humble station in an unimportant province, and died a slave's death, yet did this to redeem all men, rich and poor, freemen and slaves, citizens and barbarians, required a completely new way of looking at human beings; if all are children of God and equally capable of salvation, then all, irrespective of status or talent, vice or virtue, merit the serious attention of the poet, the novelist and the historian."

Auden made a strong claim for the impact of religion on our image of ourselves. In another place he made an equally strong claim for the importance of science: "As biological organisms made of matter, we are subject to the laws of physics and biology: as conscious persons who create our own history we are free to decide what that history shall be. Without science, we should have no notion of equality; without art, no notion of liberty."

In cultures outside Europe, religions other than Christianity have been important to the growth of civilization. Everywhere, religion and ethics are strongly coupled. The coupling between ethics and science is a major theme of this book. We may hope that groups of citizens united by ethical concerns may gain sufficient strength to shape history in the future, as they have done in the past. But ethical considerations can prevail over short-sighted self-interest only if the voice of religion is added to the voice of science. Both must be heard, if our ethical choices are to be at the same time rational and humane.

Science is a friendly international club to which I am privileged to belong. Scientists all over the world are united in a culture that gives hope of a better future for all of us. But a scientist looking out to the horizon must also try to identify the cloud no bigger than a man's hand that may grow into a deluge. The voices of both religion and science warn us that we must be vigilant. Knowledge is dangerous, as Adam and Eve learned when they tasted the forbidden fruit of the Tree of Knowledge of Good and Evil. The more we know, the greater the power we shall give to our children for good or evil, and the more responsibility we have to give them early warning of disasters.

Science is my territory, but science fiction is the landscape of my dreams. The year 1995 was the hundredth anniversary of the publication of H. G. Wells's *The Time Machine*, perhaps the darkest view of the human future ever imagined. Wells used a dramatic story to give his contemporaries a glimpse of a possible future. His purpose was not to predict but to warn. He was angry with the human species for its failures and follies. He was especially angry with the English class system under which he had personally suffered, a system that divided people into idle rich and exploited poor, the rich enjoying the refinements of art and beauty while the poor were condemned to lives of ignorance and ugliness. Wells was warning his readers, and his English upper-class readers in particular, that the gross inequality and injustice of their society was leading them to disaster. If you continue along the way you are going, his story told them, here is the way you will end, with humanity split into two species, prey and predators: the Eloi singing and dancing in the sunshine and the Morlocks keeping the machines running underground; the Eloi having lost through indolence their practical and intellectual skills, the Morlocks tending their erstwhile cousins like cattle as a convenient source of meat.

It is impossible to measure how much direct influence Wells's writings had on the social history of England. When I worked as a scientist giving technical advice to the Royal Air Force in the second world war, my chief, Reuben Smeed, formulated a rule to guide our efforts. Smeed's Rule says that you can either get something done or get the credit for it, but not both. To be effective in influencing policy or in changing society, you must make sure that people in positions of power adopt your ideas as their own. You can never know whether your personal influence was or was not decisive. In the case of Wells, we know that *The Time Machine* became an immediate best-seller, and that Wells was for many years the most widely read writer in Britain on social themes. Wells and his friends in the Fabian Society were tirelessly preaching the cause of social justice. We know that during the fifty years of Wells's working life—from the publication of *The Time Machine* in 1895 until his death in 1946—the social injustices and inequalities of English society were gradually ameliorated, as the British ruling class developed a social conscience. And we know that in the fifty years since he died, England has gradually reverted to a class system with inequalities almost as sharp as those that he fought against as a young man and lampooned in his novels. Based on the evidence, I think we may, in spite of Smeed's Rule, give Wells some credit for the improvements that occurred in English society during his lifetime.

Into *The Time Machine* Wells poured his personal anguish and his scientific detachment, his sympathetic understanding of the individual human soul and his unsympathetic understanding of the human species. He was the first novelist to place his characters, with their individual passions and personalities, within the larger framework of biological evolution. He saw the human species as a deeply flawed biological experiment, likely to fail because of internal weaknesses even if it did not succumb to external calamities. The tragic history of the twentieth century has not made Wells's vision less plausible.

The novel ends on a note of philosophical melan-

ROD TAYLOR, IN *THE TIME MACHINE*
MGM, 1960. Reproduced by permission of the Kobal Collection.

choly. After the Time Traveler's tale of horror and degeneration has been told, and he has vanished from our sight with his machine, the narrator of the story reflects upon the meaning of his voyage. "To me the future is still black and blank—is a vast ignorance, lit at a few casual places by the memory of his story. And I have by me, for my comfort, two strange white flowers—shrivelled now, and brown and flat and brittle—to witness that even when mind and strength had gone, gratitude and a mutual tenderness still lived on in the heart of man." The ending of *The Time Machine* passes beyond the violence and cataclysm of conventional science-fiction endings, just as the last scene of *King Lear* passes beyond the corpse-filled final scenes of *Macbeth* and *Hamlet* into a deeper quietness. As an artist, Wells was as prolific and almost as many-sided as Shakespeare. Like Shakespeare, he wrote tragedies and comedies and histories. Unlike Shakespeare, he began with tragedies and then went on to comedies and histories. Other disagreeable visions of the future have been written by writers less gifted than Wells, but none has equaled *The Time Machine* as a work of art.

Since Wells, we have had a hundred more years of science from which to learn, a hundred more years of history to contemplate. One lesson that we learn from science and from history is that the future is unpredictable. Despite his scientific training as a biologist, Wells never imagined the discoveries that would create the new science of molecular biology soon after his death and dominate the landscape of biology into the next millennium. In his *Outline of History* Wells labeled the symbols of nationalism "tribal gods of the nineteenth century," and did not imagine that these vestiges of tribal loyalty would endure with even greater virulence to the end of the twentieth. When Wells attempted to predict the future, as in later life he often did, he usually failed. When he imagined future worlds, using his skills as a novelist to enlarge our vision and remind us of our responsibilities, he succeeded brilliantly.

In looking toward the future, I write about those things I happen to know best—a small fraction of science and an even smaller fraction of technology. I use stories, imagined and real, to explore the interplay of science and technology with evolution and ethics. Fashionable scientific topics such as complexity and string theory, and fashionable environmental problems such as global warming and overpopulation, will be passed by in silence, not because I consider them unimportant but because I have nothing fresh to say about them. In discussions of human affairs, I turn for guidance not to sociology but to case studies and science fiction. For me, Wells's *The Time Machine* provides more insight into past and future worlds than any statistical analysis, because insight requires imagination.

chapter one

STORIES

BICYCLE SERIES I

By Janet Stern, 1992. Reproduced by permission of the artist.

SUCCESSFUL TECHNOLOGIES OFTEN BEGIN as hobbies. Jacques Cousteau invented scuba diving because he enjoyed exploring caves. The Wright brothers invented flying as a relief from the monotony of their normal business of selling and repairing bicycles. A little earlier, the bicycle and the automobile began as recreational vehicles, as means for people of leisure to explore the countryside, before smooth roads existed to make riding and driving efficient. In all these technologies, the pioneers were spending their money and risking their lives for nothing more substantial than fun. Scuba diving is fun, flying is fun, riding bicycles and driving cars are fun, especially in the early days when nobody else is doing it. Even today, when each of these four hobbies has grown into a huge industry, when legal regulations are enforced to reduce the risks as far as possible, sport and recreation are still supplying much of the motivation for pushing the technologies ahead.

The history of flying is a good example to look at in detail for insight into the interaction of technology with human affairs, because two radically different technologies were competing for survival—in the beginning they were called heavier-than-air and lighter-than-air. The airplane and the airship were not only physically different in shape and size but also sociologically different. The airplane grew out of dreams of personal adventure. The airship grew out of dreams of empire. The image in the minds of airplane-builders was a bird. The image in the minds of airship-builders was an oceanliner.

We are lucky to have a vivid picture of the creative phases of these technologies, written by a man who was deeply involved in both and was also a gifted writer, Nevil Shute Norway. Before he became the famous novelist Nevil Shute—author of *Pied Piper, A Town like Alice, On the Beach,* and other wonderful stories—he was an aeronautical engineer working professionally on the design of airplanes and airships. He wrote an autobiography with the title *Slide Rule*, describing his life as an engineer.

Norway did not start out with any bias for airplanes and against airships. He worked on both with equal dedication, and he was particularly proud of his part in the design of the airship R100. He worked on it for six years, from the moment of conception in 1924 to the delivery in 1930, and flew on its triumphant maiden voyage in 1930, from London to Montreal and back. From a technical point of view, airships then had many advantages over airplanes, and the R100 was a technical success. But Norway saw clearly that the fate of airships and airplanes did not depend on technical factors alone. Even before he became a professional writer, he was more interested in people than in nuts and bolts. He saw and recorded the human factors that made the building of airplanes fun and made the building of airships a nightmare.

After finishing the R100, Norway started a company of his own, Airspeed Limited. It was one of the hundreds of small companies that were inventing and building and selling airplanes in the 1920s and 30s. Norway estimated that 100,000 different varieties of airplane were flown during those years. All over the world, enthusiastic inventors were selling airplanes to intrepid pilots and to fledgling airlines. Many of the pilots crashed and many of the airlines became bankrupt. Out of 100,000 types of airplane, about 100 survived to form the basis of modern aviation. The evolution of the airplane was a strictly Darwinian process in which almost all the varieties of airplane failed, just as almost all species of animal become extinct. Because of the rigorous selection, the few surviving airplanes are astonishingly reliable, economical, and safe.

The Darwinian process is ruthless, because it depends upon failure. It worked well in the evolution of airplanes because the airplanes were small, the companies that built them were small, and the costs of failure in money and lives were tolerable. Planes crashed, pilots were killed, and investors were ruined, but the scale of the losses was not large enough to halt the process of evolution. After the crash, new pilots and new investors would always appear with new dreams of glory. And so the selection process continued, weeding out the unfit, until airplanes and companies had grown so large that further weeding was officially discouraged. Norway's company was one of the few that survived the weeding and became commercially profitable. As a result, it was bought out and became a division of De Havilland, losing the freedom to make its own decisions and take its own risks. Even before De Havilland took over the company, Norway decided that the business was no longer fun. He stopped building airplanes and started his new career as a novelist.

The evolution of airships was a different story, dominated by politicians rather than by inventors. British politicians in the 1920s were acutely aware that the century of world-wide British hegemony based upon sea power had come to an end. The British Empire was still the biggest in the world but could no longer rely on the Royal Navy to hold it together. Most of the leading politicians, both Conservative and Labor, still had dreams of empire. They were told by their military and political advisers that in the modern world air power was replacing sea power as the emblem of greatness. So they looked to air power as the wave of the future that would keep Britain on top of the world. And in this context it was natural for them to think of airships rather than airplanes as the vehicles of imperial authority. Airships were superficially like oceanliners, big and visually impressive. Airships could fly nonstop from one end of the empire to the other. Important politicians could fly in airships from remote dominions to meetings in London without being forced to neglect their domestic constituencies for a month. In contrast, airplanes were small, noisy, and ugly, altogether unworthy of such a lofty purpose. Airplanes at that time could not routinely fly over oceans. They could not stay aloft

L'ESPAGNE
By James Flora, 1988. Reproduced by permission of the artist.

for long and were everywhere dependent on local bases. Airplanes were useful for fighting local battles, but not for administering a worldwide empire.

One of the politicians most obsessed with airships was the Labor Peer Lord Thompson, Secretary of State for Air in the Labor governments of 1924 and 1929. Lord Thompson was the driving force behind the project to build the R101 airship at the government-owned Royal Airship Works at Cardington. Being a socialist as well as an imperialist, he insisted that the government factory get the job. But as a compromise to keep the Conservative opposition happy, he arranged for a sister ship, the R100, to be built at the same time by the private firm Vickers Limited. The R101 and R100 were to be the flagships of the British Empire in the new era. The R101, being larger, would fly nonstop from England to India and perhaps later to Australia. The R100, a more modest enterprise, would provide regular service over the Atlantic between England and Canada. Norway, from his position in the team of engineers designing the R100, had a front-seat view of the fate of both airships.

The R101 project was from the beginning driven by ideology rather than by common sense. At all costs, the R101 had to be the largest airship in the world, and at all costs it had to be ready to fly to India by a fixed date in October 1930, when Lord Thompson himself would embark on its maiden voyage to Karachi and back, returning just in time to attend an Imperial Conference in London. His dramatic arrival at the conference by airship, bearing fresh flowers from India, would demonstrate to an admiring world the greatness of Britain and the Empire, and incidentally demonstrate the superiority of socialist industry and of Lord Thompson himself. The huge size and the fixed date were a fatal combination. The technical problems of sealing enor-

R101 AT THE MOORING-MAST
1930. Reproduced by permission of the Smithsonian Institution.

mous gasbags so that they should not leak were never solved. There was no time to give the ship adequate shake-down trials before the voyage to India. It finally took off on its maiden voyage, soaking wet in foul weather, with Lord Thompson and his several thousand pounds of lordly baggage on board. The ship had barely enough lift to rise above its mooring-mast. Eight hours later it crashed and burned on a field in northern France. Of the fifty-four people on board, six survived. Lord Thompson was not among them.

Meanwhile, the R100, with Norway's help, had been built in a more reasonable manner. Its gasbags did not leak, and it had an adequate margin of lift to carry its designed pay-load. The R100 completed its maiden voyage to Montreal and back without disaster, seven weeks before the R101 left England. But Norway found the voyage far from reassuring. He reports that the R100 was violently tossed around in a local thunderstorm over Canada and was lucky to have avoided being torn apart. He did not judge it safe enough for regular passenger service. The question whether it was safe enough became moot after the R101 disaster. After one such disaster, no passengers would be likely to volunteer for another. The R100 was quietly dismantled and the pieces sold for scrap. The era of imperial airships had come to an end.

The announced purpose of the R100 was to provide a reliable passenger service between England and Canada, arriving and leaving once a week. After the airship failed, Lord Cunard, the owner of the Cunard shipping company, asked his engineers what it would take to provide a weekly service across the Atlantic using only two oceanliners. At that time it took seven or eight days for a ship to cross the Atlantic, so that a weekly service needed at least three ships. To do it with two ships would require crossing in five days, with two days margin for bad weather, loading, and unloading. The Cunard engineers designed the *Queen Mary* and the

Queen Elizabeth to cross in five days. To do this economically, because of the way wave-drag scales with speed and size, the two ships had to be substantially larger than other oceanliners. Lord Cunard felt confident that the business of transporting passengers by ship could remain profitable for a few more decades, and he ordered the ships to be built.

In due course, after the interruption caused by the second world war, they were carrying passengers profitably across the ocean and incidentally breaking speed records. The British public was proud of these ships, which regularly won the famous Blue Ribbon for the fastest Atlantic crossing. The public imagined that the ships were designed to win the Blue Ribbon, but Lord Cunard said the public misunderstood the purpose of the ships completely. He said his purpose was always to build the smallest and slowest ships that could do a regular weekly service. It was just an unfortunate accident that to do this job you had to break records. The ships continued their weekly sailings profitably for many years, until the Boeing 707 put them out of business.

☙

While oceanliners were still enjoying their heyday, before the triumph of the Boeing 707, another tragedy of ideologically driven technology occurred. This was the tragedy of the Comet jetliner. During World War II the De Havilland company had built bombers and jet fighters and acquired an appetite for bigger things. After the war, the company went ahead with the design of the Comet, a commercial jet that could fly twice as fast as the propeller-driven transport planes of that era. At the same time, the British government established the British Overseas Airways Corporation, a state-owned monopoly with responsibility for long-distance air routes. The Empire was disintegrating rapidly, but enough of it remained to inspire the planners at BOAC with new dreams of glory. Their dream was to deploy a fleet of Comets on the Empire routes that BOAC controlled, from London south to Africa and east to India and Australia.

The dream was seductive because it meant that Britain would move into the jet age five years ahead of the slow-moving Americans. While the Boeing Company hesitated, the Comets would be flying. The Comets would display to the world the superiority of British technology, and incidentally demonstrate that the Empire, now renamed the Commonwealth, was still alive. After the BOAC Comets had shown the way, other airlines all over the world would be placing orders with De Havilland. The dreams that inspired the Comet were the same dreams that inspired the R101 twenty years earlier. The heirs of Lord Thompson had learned little from his fate.

The Comet enterprise made the same mistake as the R101, pushing ahead into a difficult and demanding technology with a politically dictated time-table. The decision to rush the Comet into service in 1952 was driven by the political imperative of staying five years ahead of the Americans. One man foresaw the disaster that was coming. Nevil Shute, no longer an aeronautical engineer but a well-informed bystander, published in 1948 a novel with the title *No Highway*, which described how political pressures could push an unsafe airplane into service. The novel tells the story of a disaster that is remarkably similar to the Comet disasters that happened four years later.

The fatal flaw of the Comet was a concentration of stress at the corners of the windows. The stress caused the metal skin of the plane to crack and tear open. The cracking occurred only at high altitudes when the plane was fully pressurized. The result was a disintegration of the plane and strewing of wreckage over wide areas, leaving no clear evidence of the cause. Two planes were destroyed in this way, one over India and one over Africa, killing everybody on board. After the second crash, the Comets stopped flying. For five years no jetliners flew, until the Americans were ready with their reliable and thoroughly tested Boeing 707. It took a hundred deaths to stop the Comets from flying, twice as many as it took to stop the airships. If the Secretary of State for Air had been on board the first Comet when it crashed, the second crash might not have been necessary.

Nevil Shute explains how it happened that the R101 and the Comets were allowed to carry passengers without adequate flight-testing. It happened because of a clash of two cultures, the culture of politics and the culture of engineering. Politicians were making crucial decisions about technical matters which they did not understand. The job of a senior politician is to make decisions. Political decisions are often made on the basis of inadequate knowledge, and usually without doing much harm. In the culture of politics, a leader gains respect by saying: "The buck stops here." To take a chance of making a bad decision is better than to be indecisive. The culture of engineering is different. An engineer gains respect by saying: "Better safe than sorry." Engineers are trained to look for weak points in a design—to warn of potential disaster. When politicians are in charge of an engineering venture, the two cultures clash. When the venture involves machines that fly in the air, a clash tends to result in a crash.

Aviation is the branch of engineering that is least

forgiving of mistakes. But from a wider point of view, unforgivingness may be a virtue. In the long view of history, the victims of the R101 and the Comets did not die in vain. They left as the legacy of their tragedy the extraordinarily safe and reliable airplanes that now fly every day across continents and oceans all over the world. Without the harsh lessons of disaster and death, the modern jetliner would not have evolved.

My friend Albert Hirschman has found other places where unforgivingness is a virtue. He is an economist who has spent much of his life studying Latin American societies and giving advice to their governments. He has also given advice to newly independent countries in Africa. He is often asked by the leaders of poor countries, "Should we put our limited resources into roads or into airlines?" When this question is asked, the natural impulse of an economist is to say "roads," because the money spent on roads provides jobs for local people, and the roads benefit all classes of society. In contrast, the building of a national airline requires the import of foreign technology, and the airline benefits only the minority of citizens who can afford to use it. Nevertheless, long experience in Africa and Latin America has taught Hirschman that "roads" is usually the wrong answer. In the real world, roads have several disadvantages. The money assigned to road-building tends to fall into the hands of corrupt local officials. Roads are easier to build than to maintain. And when, as usually happens, the new roads decay after a few years, the decay is gradual and does not create a major scandal. The end-result of road-building is that life continues as before. The economist who said "roads" has achieved little except a small increase in the wealth and power of local officials.

Contrast this with the real-world effect of building a national airline. After the money is spent, the country is left with some expensive airplanes, some expensive airports, and some expensive modern equipment. The foreign technicians have left and local people must be trained to operate the system. Unlike roads, airplanes do not decay gracefully. A crash of an airliner is a highly visible event and brings unacceptable loss of prestige to the rulers of the country. The victims tend to be people of wealth and influence, and their deaths do not pass unnoticed. The rulers have no choice. Once they own an airline, they are compelled to see to it that the airline is competently run. They are forced to create a cadre of highly motivated people who maintain the machines, come to work on time, and take pride in their technical skills. As a result, the airline brings to the country indirect benefits that are larger than its direct economic value. It creates a substantial body of citizens accustomed to strict industrial discipline and imbued with a modern work ethic. And these citizens will in time find other useful things to do with their skills besides taking care of airplanes. In this paradoxical way, the unforgivingness of aviation makes it the best school for teaching a traditional society how to modernize.

This is not the first time that an unforgiving technology has transformed the world and forced traditional societies to change. The role of aviation today is similar to the role of sailing ships in the preindustrial world. King Henry VIII of England, the most brutal and most intelligent of English monarchs, destroyer of monasteries and founder of colleges, murderer of wives and composer of madrigals, for whose soul regular prayers are still said at Trinity College Cambridge in gratitude for his largesse, understood that the most effective tool for modernizing England was the creation of a Royal Navy. It was not by accident that the industrial revolution of the eighteenth century began in England, in the island where daily life and economics had been dominated for 300 years by the culture of sailing ships. When the young Tsar Peter the Great of Russia, a kindred spirit to Henry, decided that the time had come to modernize the Russian empire, he prepared himself for the job by going to work as an apprentice in a shipyard.

❧

The R101 and Comet tragedies are examples of the baleful effects of ideology, the ideology in those cases being old-fashioned British imperialism. Today, the British Empire is ancient history, and its ideology is dead. But technologies driven by ideology are likely to run into trouble, even when the ideology is not so outmoded. Another powerful ideology that ran into trouble is nuclear energy. All over the world, after the end of World War II, the ideology of nuclear energy flourished, driven by an intense desire to create something peaceful and useful out of the ruins of Hiroshima and Nagasaki. Scientists and politicians and industrial leaders were equally bewitched by this vision, that the great new force of nature that killed and maimed in war would now make deserts bloom in peace. Nuclear energy was so strange and powerful that it looked like magic. It was easy to believe that this magic could bring wealth and prosperity to poor people all over the earth. So it happened that in all large countries and in many small ones, in democracies and dictatorships, in communist and capitalist societies alike, Atomic Energy Authorities were created to oversee the miracles that nuclear energy was expected to perform. Huge funds were poured into nuclear laboratories in the confident belief that these were sound investments for the future.

I visited Harwell, the main British nuclear research establishment, during the early days of nuclear enthusiasm. The first director of Harwell was Sir John Cockcroft, a first-rate scientist and an honest public servant. I walked around the site with Cockcroft, and we looked up at the massive electric power lines running out of the plant, over our heads and away into the distance. Cockcroft remarked, "The public imagines that the electricity is flowing out of this place into the national grid. When I tell them that it is all flowing the other way, they don't believe me."

There was nothing wrong, and there is still nothing wrong, with using nuclear energy to make electricity. But the rules of the game must be fair, so that nuclear energy competes with other sources of energy and is allowed to fail if it does badly. So long as it is allowed to fail, nuclear energy can do no great harm. But the characteristic feature of an ideologically driven technology is that it is not allowed to fail. And that is why nuclear energy got into trouble. The ideology said that nuclear energy must win. The promoters of nuclear energy believed as a matter of faith that it would be safe and clean and cheap and a blessing to humanity. When evidence to the contrary emerged, the promoters found

STANROCK TAILINGS WALL

The wall of white sand in back of the trees is made up of radioactive mill wastes from uranium mining in the Elliot Lake region of Ontario. Photograph by Robert del Tredici, 1986. Reproduced by permission of the photographer.

ways to ignore the evidence. They wrote the rules of the game so that nuclear energy could not lose. The rules for cost-accounting were written so that the cost of nuclear electricity did not include the huge public investments that had been made to develop the technology and to manufacture the fuel. The rules for reactor safety were written so that the type of light-water reactor originally developed by the United States Navy for propelling submarines was by definition safe. The rules for environmental cleanliness were written so that the ultimate disposal of spent fuel and worn-out machinery was left out of consideration. With the rules so written, nuclear energy confirmed the beliefs of its promoters. According to these rules, nuclear energy was indeed cheap and clean and safe.

The people who wrote the rules did not intend to deceive the public. They deceived themselves, and then fell into a habit of suppressing evidence that contradicted their firmly held beliefs. In the end, the ideology of nuclear energy collapsed because the technology that was not allowed to fail was obviously failing. In spite of the government subsidies, nuclear electricity did not become significantly cheaper than electricity made by burning coal and oil. In spite of the declared safety of light-water reactors, accidents occasionally happened. In spite of the environmental advantages of nuclear power plants, disposal of waste fuel remained an unsolved problem. The public, in the end, reacted harshly against nuclear power because obvious facts contradicted the claims of the promoters.

When a technology is allowed to fail in competition with other technologies, the failure is a part of the normal Darwinian process of evolution, leading to improvements and possible later success. When a technology is not allowed to fail, and still it fails, the fail-

THE BECHEREL REINDEER

This radioactive meat in a slaughterhouse freezer in Swedish Lapland came from reindeer that fed on lichen contaminated with cesium-137 from the Chernobyl cloud. Photograph by Robert del Tredici, 1986. Reproduced by permission of the photographer.

ure is far more damaging. If nuclear power had been allowed to fail at the beginning, it might well have evolved by now into a better technology which the public would trust and support. There is nothing in the laws of nature that stops us from building better nuclear power plants. We are stopped by deep and justified public distrust. The public distrusts the experts because they claimed to be infallible. The public knows that human beings are fallible. Only people blinded by ideology fall into the trap of believing in their own infallibility.

The tragedy of nuclear fission energy is now almost at an end, so far as the United States is concerned. Nobody wants to build any new fission power plants. But another tragedy is still being played out, the tragedy of nuclear fusion. The promoters of fusion are making the same mistakes that the promoters of fission made thirty years earlier. The promoters are no longer experimenting with a variety of fusion schemes in order to evolve a machine that might win in the marketplace. They long ago decided to concentrate their main effort upon a single device, the Tokamak, which is declared by ideological fiat to be the energy producer for the twenty-first century. The Tokamak was invented in Russia, and its inventors gave it a name that transliterates euphoniously into other languages. All the countries with serious programs of fusion research have built Tokamaks. One of the largest and most expensive is in Princeton. To me it looks like a plumber's nightmare, a dense conglomeration of pipes and coils with no space for anybody to go in and fix it when it needs repairs. But the people who built it believe sincerely that it is an answer to human needs. The various national fusion programs are supposed to converge upon a huge international Tokamak, costing many billions of

THE TOKAMAK FUSION TEST REACTOR

1993. Reproduced by permission of the Plasma Physics Laboratory, Princeton University.

dollars, which will be the prototype for the fusion power producers of the future. The usual claims are made, that fusion power will be safe and clean, although even the promoters are no longer saying that it will be cheap. The existing fusion programs have stopped the evolution of a new technology that might actually fulfill the hopes of the promoters. What the world needs is a small, compact, flexible fusion technology that could make electricity where and when it is needed. The existing fusion program is leading to a huge source of centralized power, at a price that nobody except a government can afford. It is likely that the existing fusion program will sooner or later collapse as the fission program collapsed, and we can only hope that some more useful form of fusion technology will rise from the wreckage.

My last story about a technology driven by ideology is the story of ice ponds. Ted Taylor was the chief promoter of ice ponds. Of all my friends, he is the one who best combines technical inventiveness with high moral principles. In his youth he was a designer of nuclear weapons at Los Alamos. He later worked in the Department of Defense in Washington, with responsibility for safeguarding nuclear weapon stockpiles. After this exposure to the realities of nuclear-industrial politics, he became an antinuclear activist, resigned from government service, and campaigned publicly for better safeguards against theft of plutonium and other nuclear materials. He decided to devote the rest of his life to developing alternative technologies to replace nuclear energy. The search for a sustainable and environmentally benign source of energy led him to ice ponds.

Ice ponds could be a clean source of energy for refrigeration, in any region where winter night-time temperatures fall below freezing for at least ten nights per year. The idea of the ice pond is to store a large volume of snow for half a year, so that the snow can be made in winter and used for refrigeration in summer. The snow is made in winter by squirting water through a fog nozzle of the kind used by firefighters to produce a fine spray. Provided that the air temperature is below freezing, the spray falls to the ground as snow or slush which accumulates in the pond. The pile of snow is covered with an insulating blanket. The pond is connected by water pipes with the building to be refrigerated. In summer, cold water is extracted from the bottom of the pond and warm water is returned to the top. If the pond is big and deep enough, the snow will last through the summer and the building will stay cool. The energy required to make the snow and pump the circulating water is far less than the energy required for conventional electric refrigeration.

Taylor dreamed of harnessing the natural cycle of the seasons to replace electricity generated in polluting power stations. Inspired by this dream, he built a demonstration ice pond at Princeton University and used it successfully to refrigerate a small building. He persuaded the Prudential Insurance Company to install a larger ice pond to provide air-conditioning for a larger building. He persuaded the Kutter cheese company in

THE PRINCETON ICE POND

*1981. Reproduced by permission of the Communications Office,
Princeton University.*

New York State to build ice ponds to refrigerate a
cheese factory. He persuaded the village of Greenport
on Long Island to build an ice pond for the purpose of
purifying sea water. The Greenport ice pond used salt
water from the Atlantic Ocean to make snow. Within a
few weeks, the salt drained out from the bottom of the
snow pile, and the bulk of the remaining snow was pure
enough to satisfy the New York State quality standards
for drinking water.

These projects were all one-of-a-kind, designed to
explore the technology of ice ponds. They met with

varying degrees of success. The Princeton University ice pond worked well, but it was dismantled after two years because it served no useful purpose except to demonstrate feasibility. The Prudential ice pond never worked satisfactorily, because the operation was taken out of Taylor's hands and entrusted to Prudential employees who did not understand the system. The Greenport ice pond worked well technically, but it became entangled in local political quarrels. It was never connected to the Greenport water supply and was finally dismantled by majority vote of the village government. The Kutter ice ponds are the only ones that survived and continue to perform a useful service for their owners.

Taylor had hoped, after these demonstration projects had shown the world what ice ponds could do, to make ice ponds into a commercial product. He hoped to start a profit-making business, manufacturing ice ponds and selling them to the public. The buyers that he hoped to attract were owners of food-processing plants, owners of commercial buildings, and real estate developers building groups of townhouses. The buyers would be saving money by using less electricity, and at the same time acquiring a visible symbol of environmental virtue. The ice pond beside a building, like the collector of solar energy on the roof, would quietly assert the owner's concern for the balance of nature. Unfortunately, Taylor never found his buyers, and ice ponds never became a standardized product that could be delivered to the customer for a fixed price.

Why did the ice pond venture fail? There were many reasons for the failure. Taylor underestimated the practical difficulties of operating heavy equipment in winter out-of-doors. The job of covering a huge pile of snow with an insulating blanket turned out to be unexpectedly difficult. If the blanket was not properly laid, the snow would not last through the summer. The failure of the Prudential ice pond was mainly due to the failure of the blanket. Maintaining an ice pond required constant attention. The Kutter ice ponds succeeded because the Kutter brothers who owned the cheese factory were enthusiastic tinkerers, happy to spend long hours taking care of the machinery when anything went wrong. For less adventurous owners who did not enjoy working out-of-doors in foul weather, the problems of maintenance would have been a constant headache. These practical deficiencies of the ice pond made any commercial success unlikely.

But there was a more fundamental reason for the failure of Taylor's dream. There was never a market pull for ice ponds; there was only an ideological push.

Successful technologies are pulled along by the needs of the buyers, not pushed along by the ideology of the sellers. Taylor violated the first rule for running a successful business: "Know your market."

The ice pond, like the airship and the nuclear power station, is an example of a technology that was driven by ideology and failed. But the failure of ice ponds was not a tragedy like the failures of airships and nuclear power. Ice ponds failed quickly and caused minimal losses to society. Little money and no human lives were wasted. Even Taylor himself emerged from the debacle with his spirit intact, ready to start new ventures. His love affair with ice ponds has left behind a legacy of useful knowledge. The technology of ice ponds remains as a possibility for the future. One day, perhaps, a more astute reincarnation of Taylor will find a way to turn ice ponds into a convenient and user-friendly package that will fulfill Taylor's hopes.

The moral of the ice pond story is that ideology-driven technology need not lead to disaster. It will lead to disaster only if it is protected from competition. Provided that a technology is exposed to the Darwinian process of selection, it does not matter whether it was originally motivated by profit seeking or by ideology. Ideology push may be a positive force for good, if it leads to environmentally benign technologies that can be tested in the marketplace. I do not regret the happy days that I spent with Ted Taylor and his students, helping to build the Princeton ice pond. We were luckier than the builders of airships and nuclear power plants, because we were allowed to fail.

SCIENCE

LE SENS DES RÉALITÉS

IF WE ARE LOOKING FOR NEW DIRECTIONS in science, we must look for scientific revolutions. When no scientific revolution is under way, science continues to move ahead along old directions. It is impossible to predict scientific revolutions, but it may sometimes be possible to imagine a revolution before it happens.

There are two kinds of scientific revolutions, those driven by new tools and those driven by new concepts. Thomas Kuhn in his famous book, *The Structure of Scientific Revolutions,* talked almost exclusively about concepts and hardly at all about tools. His idea of a scientific revolution is based on a single example, the revolution in theoretical physics that occurred in the 1920s with the advent of quantum mechanics. This was a prime example of a concept-driven revolution. Kuhn's book was so brilliantly written that it became an instant classic. It misled a whole generation of students and historians of science into believing that all scientific revolutions are concept-driven. The concept-driven revolutions are the ones that attract the most attention and have the greatest impact on the public awareness of science, but in fact they are comparatively rare. In the last 500 years, in addition to the quantum-mechanical revolution that Kuhn took as his model, we have had six major concept-driven revolutions, associated with the names of Copernicus, Newton, Darwin, Maxwell, Freud, and Einstein. During the same period there have been about twenty tool-driven revolutions, not so impressive to the general public but of equal importance to the progress of science. Two prime examples of tool-driven revolutions are the Galilean revolution resulting from the use of the telescope in astronomy, and the Crick-Watson revolution resulting from the use of X-ray diffraction to determine the structure of big molecules in biology.

The effect of a concept-driven revolution is to explain old things in new ways. The effect of a tool-driven revolution is to discover new things that have to be explained. In almost every branch of science, and especially in biology and astronomy, there has been a preponderance of tool-driven revolutions. We have been more successful in discovering new things than in explaining old ones. In recent times my own field of physics has had great success in creating new tools that have started revolutions in biology and astronomy. Physics has been less successful in creating new concepts with which to understand its own discoveries.

After the Crick-Watson revolution of the 1950s, the next great tool-driven revolution was the advent of electronic computers and memory banks in the 1960s. Electronic data processing revolutionized every branch of experimental science, while electronic computer simulations revolutionized every branch of theoretical science. Both the Crick-Watson and the computer revolutions were driven by tools imported from physics.

The computer incidentally caused a revolution in physics itself, increasing the power of physical theories to interpret experiments and predict phenomena. The computer is a prime example of an intellectual tool. It is not a concept but a tool for clear thinking. It helps us to think more clearly by enabling us to calculate more precisely. The computer has also had a revolutionary effect in narrowing the gap between modern mathematics and theoretical physics. Sophisticated mathematical ideas that were alien to physicists of an earlier generation are now routinely used to build physical theories.

The computer revolution began with a new tool and quickly emerged into a new style as computers became small, cheap, and ubiquitous. At the beginning of the revolution, when John von Neumann built his computer at Princeton, computers were large and expensive. Von Neumann's computer was dedicated to two big projects, weather prediction in daytime and hydrogen bombs at night. Twenty years later, when the revolution was in full swing, computers had become personal, available to anyone who needed them, and applicable to a huge variety of purposes. Centralized management was displaced by do-it-yourself improvisation. Computers are now as common as dishwashers and are used indiscriminately as tools or toys.

It often happens that a scientific revolution is accompanied by a change in style. I like to use the names of Napoleon and Tolstoy to symbolize two contrasting styles: rigid organization and discipline represented by Napoleon, creative chaos and freedom represented by Tolstoy. In the world of computers, Napoleon is the massive IBM main-frame; Tolstoy is the humble Macintosh. The computer revolution was an escape from the Napoleonic ambitions of von Neumann to the

Tolstoyan anarchy of the Internet. Future revolutions will bring more such escapes.

⚜

In the last few years I had the opportunity of visiting two extraordinary laboratories, both in Switzerland: the international particle physics laboratory known as CERN near Geneva, and the IBM research center at Rüschlikon near Zürich. These happen to be the two places where the most spectacular discoveries in the physics of the last twenty years have been made, at CERN in particle physics and at Rüschlikon in condensed-matter physics. The people at CERN discovered the beautiful world of W and Z particles; the people at Rüschlikon discovered scanning tunneling microscopes and high-temperature superconductors.

These two laboratories are good examples of the two styles of science. CERN with its big machines and its centralized administration belongs firmly to the Napoleonic tradition, even if the new Director-General, Christopher Llewellyn-Smith, does not wear the Imperial tiara as flamboyantly as his predecessor, Carlo Rubbia. The IBM laboratory at Rüschlikon is just as firmly Tolstoyan, with a social structure resembling an extended family, and nobody giving orders. The two styles are appropriate to the different tasks that the two laboratories are engaged in. To operate successfully a ma-

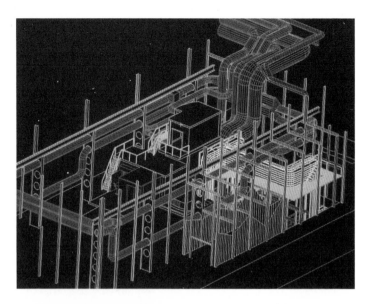

**THE VENTILATION SYSTEM, CERN RESEARCH
CENTRE, GENEVA**

*Computer-aided design (CAD) is invaluable in designing the
optimal layout of such systems, because changes to the design may
be effected instantly and the impact on the overall facility may be
quickly determined. Photograph by David Parker, 1985.
Reproduced by permission of David Parker/Science Photo Library.*

chine of the size and complexity of LEP, the Large
Electron-Positron Collider, Napoleonic centralization is
unavoidable. Big machines require big egos to build and
operate them. Each experiment at LEP resembles a
military campaign, with elaborate logistics and time-
tables prepared several years in advance. On the other
hand, military time-tables would have been totally out
of place at Rüschlikon, where the major discoveries

were unexpected and unplanned. At Rüschlikon, the administration provides excellent equipment for talented scientists to play with, and then gives them freedom to play.

In all areas of science the future will bring opportunities to build new tools and make new discoveries, some requiring Napoleonic discipline, others requiring Tolstoyan freedom. Both on the national and the international level, the funding for science is likely to be unstable. Money for science will be increasingly spasmodic and unpredictable. In such an environment, Napoleonic enterprises will be ill-adapted to survive; Tolstoyan enterprises will do better. We should be prepared to shift science as far as we can toward a Tolstoyan style of operation. In some sciences such as microbiology and neurobiology, Tolstoyan chaos already prevails to a large extent. The two areas of science for which a shift away from Napoleonic rigidity will be most difficult are particle physics and space science. These happen also to be the areas of my own professional activity as a scientist.

When I began life as a particle physicist fifty years ago, most of the major discoveries were made in Europe by people studying the cosmic rays that bombard the earth from outer space. Particle physics was done by observing the debris produced by cosmic rays as they pass through the atmosphere and through the experimental apparatus. The debris consists of particles with short lifetimes and unfamiliar names. Particle physics was then in a Tolstoyan phase. Three young Italians, Conversi, Pancini, and Piccioni, working with homemade particle counters in the chaos of postwar Italy, discovered that the common cosmic ray particle, later called the muon, had only weak interactions with matter. Cecil Powell, working with microscopes and photographic plates at Bristol in England, discovered the rarer strongly interacting cosmic ray particle, which he called the pion. Other strange new particles were discovered by Rochester and Butler using old-fashioned cosmic ray cloud-chambers in Manchester. The new wave of particle physics grew in a few short years out of improvised experiments observing whatever particles nature happened to provide.

Meanwhile, the Americans at Berkeley and Cornell and Chicago were preparing to build accelerators that would produce particles in great abundance and put the European cosmic ray experimenters out of business. Within five years, the accelerators triumphed and particle physics entered a long Napoleonic phase. For forty years the accelerators grew bigger and more expensive, and the accelerator Napoleons collected their Nobel prizes for leading successively larger and more expensive teams of scientists to victory.

Now particle physics in the United States is struggling to survive the disaster of the Superconducting Supercollider. The Supercollider was a gigantic particle accelerator project that was canceled in 1993 after about 3 billion dollars had already been spent on it. The cancellation was a personal tragedy for many of my friends who had devoted the best years of their lives to the project. But when I speak of the disaster of the Superconducting Supercollider, I do not mean the cancellation of the project. The disaster occurred five years earlier, when the promoters of the Supercollider led the Congress and the public to believe that we could not do any important particle physics for less than 5 billion dollars. This belief was not only false but immensely harmful to the future of science. The harm would have been even greater if the Supercollider project had been allowed to continue. Every big new machine is a gamble. If the gamble succeeds, the machine produces important new discoveries. If the gamble fails, the machine is a waste of money and time. A gamble is reasonable if you can afford to lose without being ruined. The Supercollider was an unreasonable gamble because we could not afford to lose.

Now that the Supercollider is dead, we have a chance to recover from the disaster. Experimenters using older accelerators are demonstrating that there is plenty of good particle physics to be done with machines costing less than a billion dollars. In the meantime, the Large Hadron Collider will be continuing the Napoleonic tradition at CERN on a scale which I find reasonable and appropriate. The Large Hadron Collider will do the same kind of experiments that the Supercollider would have done, at about one fifth of the cost. If nature is unkind and does not provide important new things for the Large Hadron Collider to discover, the Europeans will have lost their gamble, but European science will not be ruined.

The time may now be ripe for a new Tolstoyan phase in particle physics. Even the scientists who believed that the Superconducting Supercollider was a splendid idea were aware that the big particle accelerators were running into a law of diminishing returns. Each step upward in the size of accelerators cost more money than earlier steps and took a longer time to produce new discoveries. The ratio of scientific output to financial input was diminishing rapidly as the size increased. At some point not far in the future, the further growth of accelerators was bound to grind to a halt.

In the meantime, nature continues to provide a free

supply of cosmic rays, and there are reasons to believe that nature may also provide a free supply of unknown and exotic particles pervading the universe. Groups of scientists in many countries are building large detectors deep underground, with the aim of studying the output of known particles from the sun and incidentally searching for exotic particles. The underground detectors observe natural events with technically sophisticated methods similar to those used for detecting particles in accelerator experiments.

Six modern underground detectors are now in operation, two in the United States, two in Japan, one in Italy, and one in Russia. Many more are being built and planned. The new generation of underground detectors has four virtues. First, they are cheaper than accelerators by about a factor of ten. Second, they are flexible and can be reprogrammed easily to search for new phenomena in response to new scientific ideas. Third, they are more likely than accelerator experiments to make unexpected or unimagined discoveries. Fourth, they lend themselves better than accelerators to an informal and Tolstoyan style of operation.

Accelerators still have virtues that underground detectors lack: high precision of measurement, high abundance of particles, and reproducibility of results. In the future, if we are lucky, radically new techniques of acceleration may allow a radical decrease in size and cost for an accelerator of given energy. My hope for the future of accelerators is a revolution driven by the new tool of laser technology. Almost all accelerators, from the first cyclotron in the 1930s to the supercollider today, have used radiofrequency electromagnetic fields to accelerate particles. The basic technique of acceleration has not changed in sixty years. And the accelerating field-strengths obtainable at radio frequencies seem to be limited to about a hundred million volts per meter. The technology of lasers, which are coherent sources of light, allows us to generate field-strengths a thousand times larger, of the order of 100 billion volts per meter. The problem of using such intense laser fields at optical frequencies to accelerate particles efficiently has not been solved. If this problem could be solved, then high-energy accelerators of a new type could be built, enormously smaller and cheaper than existing accelerators of comparable performance.

There are many obstacles to be overcome before this dream could come true, and leading experts in accelerator design have declared it to be impossible. But then, before any revolution in technology occurs, the experts in the existing technology always declare the revolution to be impossible. Otherwise it would not be a revolution. So there is still a good chance that radically improved accelerators will be invented. Even if laser acceleration does not fulfill my hopes, a revolution in particle acceleration is likely to appear from some other unexpected direction. Particle physics will continue to need accelerators as well as underground detectors. But the balance in the future is likely to shift toward underground detectors, and this is a hopeful trend for people like me who are incurable Tolstoyans. I see a bright future for particle physics after the Napoleonic era ends.

In space science, even more than in particle physics, the Napoleonic style has been dominant for the last thirty years. Missions on a grand scale, such as the *Voyager* explorations of the outer planets and the Hubble Space Telescope explorations of distant galaxies, have brought back to Earth a wealth of scientific knowledge, and also brought back political glory to the bureaucratic Napoleons of the National Aeronautics and Space Administration (NASA). But within NASA, just as in the community of particle physicists, winds of change are blowing. Space scientists are keenly aware that times are changing. Billion-dollar missions are no longer in style. Funding in the future will be chancy. The best chances of flying will go to missions that are small and cheap.

I recently spent some weeks at the Jet Propulsion

Laboratory in California. JPL built and operated the *Voyager* missions. It is the most independent and the most imaginative part of NASA. I was particularly interested in three planetary missions that the people at JPL hope to fly, *Cassini, Pluto Express,* and *Neptune Orbiter. Cassini* is the last of the grand missions belonging to the Napoleonic tradition. The *Cassini* hardware stands in a clean room at JPL, waiting for launch in 1997 if all goes well. After a long and complicated journey, *Cassini* will stay in orbit around Saturn and make close encounters with Saturn's rings and satellites, sending back to Earth far more information than the *Voyagers* could provide during their brief fly-by.

Cassini looks like *Voyager.* It is a massive spacecraft, carrying appendages on which a variety of instruments are deployed. The cost of the mission, including five years of operation in the Saturnian system, is estimated at 3.5 billion dollars. From a scientific point of view, *Cassini* is a superb mission, but from a political point of view it is highly vulnerable. It is a prime example of the sort of mission that reformers in NASA and in Congress would like to abolish.

Pluto Express and *Neptune Orbiter* are not standing in clean rooms at JPL. Both missions exist as ideas in the minds of JPL designers. *Pluto Express* would complete the *Voyager* exploration of the outer planets by taking

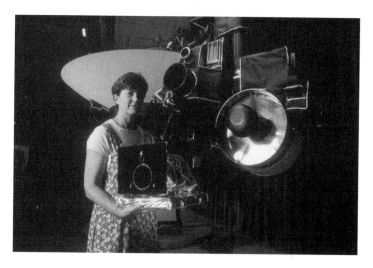

PLUTO EXPRESS INSTRUMENTS

Shown in the hands of Patricia M. Beauchamp, a co-leader of the study team at the Jet Propulsion Laboratory. The 1970s Voyager instruments, seen at the right, weigh 40 times as much and are less sensitive. Photograph by Z. Aronovsky, 1995. Reproduced by permission of Z. Aronovsky / Zuma.

high-resolution pictures in many wave-bands of Pluto and its large satellite Charon. After passing by Pluto, it would go on to explore the Kuiper belt of newly discovered smaller objects orbiting the Sun beyond the orbit of Neptune. The design of *Pluto Express* is based on a radical shrinkage of the instruments that were carried by *Voyager.* I held in my hands the prototype package of instruments for the new mission. The package weighs seven kilograms. It does the same job as the *Voyager* instruments which weighed more than a hundred kilograms. All the hardware components—optical, mechanical, structural, and electronic—have been drastically reduced in size and weight without sacrifice of performance.

Neptune Orbiter is an even more adventurous idea. It is not yet an approved proposal, much less an approved mission. It is designed not just to fly by Neptune but to go into orbit around Neptune and explore Neptune and its satellites in detail. Neptune is three times as far away as Saturn, and the idea is to do the same job at Neptune that *Cassini* will do at Saturn, but to reduce the cost by a factor of ten by using new technology.

Four new technologies are crucial for the reduction of costs. The first is the same for *Neptune Orbiter* as for *Pluto Express;* the other three are unique to *Neptune Orbiter.* They are: (1) shrinkage of instruments and computers without loss of performance; (2) atmospheric braking in the thin stratosphere of Neptune to reduce the speed of the spacecraft and achieve an orbit around the planet without using large quantities of propellant; (3) solar-electric propulsion; (4) inflatable structures deployed in space.

Solar-electric propulsion means using solar energy to generate electricity and using the electricity to drive an ion-jet engine. The ion-jet is made of xenon, a heavy gas which can be conveniently compressed to the density of water and carried in a small tank without refrigeration. The prototype xenon-ion engine was undergoing endurance tests in a tank at JPL when I visited. It must run reliably for eighteen months without loss of performance before it can be seriously considered for an operational mission. The power source for the mission is a pair of large and extremely light solar panels. The panels are large enough to provide power for housekeeping and communication with Earth even as far from the Sun as Neptune.

Inflatable structures have been extensively used on the ground as cheap covering for tennis courts and playgrounds. In space they would be used to create long light-weight beams, membranes, and dishes. They would first be blown up like big balloons by injecting small quantities of gas at low pressure, then chemically cured so that their shapes become rigid and permanent. I visited the inflatable structures group at JPL. They have dreams of reducing the costs of all large thin structures such as radio antennas, solar collectors, optical mirrors, and air brakes by a factor of a hundred. The four new technologies—miniaturized instruments, atmospheric braking, solar-electric propulsion, and inflatable structures—are sprouting at JPL in a Tolstoyan fashion, driven by enthusiasm from the bottom rather than by management from the top.

Neptune Orbiter is a daring venture, breaking new ground in many directions. It demands new technology and a new style of management. It may fail, if the managers are not as daring as the designers. The managers may transform it into a mission that is not cheap enough to fly. But solar-electric propulsion has opened the door to a new generation of cost-effective small spacecraft, taking full advantage of the enormous progress that has been made in the last thirty years in miniaturizing instruments and computers. The use of solar-electric propulsion will change the nature and style of planetary missions. Spacecraft using solar-electric propulsion may wander around the solar system, changing their trajectories from time to time to follow the changing needs of science. Solar-electric propulsion will give them the flexibility to manage their affairs in a Tolstoyan rather than a Napoleonic fashion. If *Neptune Orbiter* fails to fly, some other more daring mission will succeed. In space science, just as in particle physics, the collapse of the old Napoleonic order opens new opportunities for adventurous spirits.

Ground-based astronomy is also in the midst of a revolution. The revolution is driven by a new tool of observation, the Charge Coupled Device, popularly known as the CCD. The CCD is the detector that made modern television cameras possible. It can collect light on a million spots simultaneously and read out the pattern of light intensities directly into a computer. The CCD is causing a radical transformation of optical astronomy. Instead of calling this revolution the CCD revolution, I prefer to call it the Digital Astronomy revolution. It was predicted by the Swiss astronomer Fritz Zwicky, long before the CCD was invented. I quote from the Halley Lecture given by Zwicky in 1948 at Oxford with the title "Morphological Astronomy." Zwicky was describing a television camera system called the photoelectronic telescope that he had been working on with his friend Vladimir Zworykin at the Radio Corporation of America. "The photoelectronic telescope introduces the following new features. (1) Electrons can be accelerated from the image surface to the recording surface and power can be fed into the telescope to increase the intensity of the signals. (2) Uniform background of light may be eliminated by electric compensation. The sky background may thus be scanned away. (3) Although the original image may move, dance or scintillate because of the unsteadiness of the atmosphere, the refocused image on the recording surface can be steadied. Zworykin has actually built such an image stabilizer. (4) Automatic guiding of a telescope may be accomplished. (5) Images from photoelectronic telescopes can be televised, and the search for novae, supernovae, variable stars, comets, meteors, etc., can be put on a mass production scale, even if the telescopes are of relatively limited definition and power."

Zworykin had been a pioneer in the development of television. He lived in the house next door to me in Princeton, and was as brilliant and as cantankerous as Zwicky. Zwicky was hoping in 1948 that all the good things he mentioned in his lecture could be achieved by the Zworykin camera. The Zworykin camera did not fulfill Zwicky's hopes, but now the CCD does everything that he wanted. The main reason why the Zworykin system failed was that it still depended on photographic plates for recording images. The main reason why the CCD succeeds is that it is coupled to a digital memory instead of to a chemical image on a plate. The digital astronomy revolution had to wait until the technology of image-processing had matured, with powerful microprocessors and digital memories to match the abundance of data that the CCD could supply. Astronomy is now an intimate symbiosis of three cultures, the old culture of optical telescopes, the newer culture of electronics, and the newest culture of software engineering.

The digital astronomy revolution involves both professional scientists and amateur astronomers. A good example of digital astronomy at the professional level is the Sloan Digital Sky Survey, a project in which many of my colleagues at Princeton are actively engaged. The Sloan Survey is a modern version of the Palomar Sky Survey, the photographic survey of the northern sky which was finished in 1956 and supplied astronomers with their first accurate large-scale map of the universe. The Palomar Sky Survey plates have been enormously useful but are now about to be superseded by something better. The output of the Sloan Survey will be a photometrically precise map of the sky in five colors, plus a collection of spectra providing red shifts of about a million galaxies and other interesting objects. One byproduct of this output will be a three-dimensional view of the large-scale structure of the universe over a volume a hundred times as large as the volume covered by existing surveys.

The output of the survey will be transmitted at electronic speed to any astronomical center possessing a digital memory large enough to swallow it. The size of memory required will be measured in tens of millions of megabytes. For customers lacking such a gargantuan memory, various predigested versions of the output will be provided, with the photometric data compressed into star catalogues and galaxy catalogues supplemented by images of particularly interesting local areas. The essential difference between the Sloan Survey and previous surveys is that the output will be quantitative, consisting of precise numerical data instead of fuzzy marks on a photographic plate. The output will be packaged so that all the tricks of modern data processing can be applied to it. The purpose of the survey is to use modern digital technology to give everybody a clearer view of the universe.

The Sloan Survey is a collaborative project in which Princeton is one of seven partners. It uses a new 2.5-meter wide-field telescope, built at Apache Point, New Mexico, and dedicated to the project for five years. With luck, the survey will be finished by the year 2000. In the focal plane of the telescope there is a large array of CCD detectors. The hardware components of the project do not stretch the state of the art in telescope or detector design. The main novelty of the project lies in the software, which has to control the sequence of operations, calibrate the CCD detectors, monitor the sky quality, and apply several levels of data compression to the output before distributing it to the users. The major share of the cost of the project is paid by the Sloan Foundation, following the good example of the National Geographical Society which funded the Palomar Sky Survey fifty years earlier. The total cost is estimated to be 14 million dollars, including the capital cost of the telescope. This is about a fifth of the cost of a major ground-based observatory, about a hundredth of the cost of the Hubble Space Telescope.

After our little survey is finished, there will be other surveys putting into digital memory larger and deeper maps of the universe. There are many directions for future surveys to explore. One survey may push toward fainter and more distant objects, another toward higher angular resolution, another into a wider choice of wavelengths, another into higher spectral resolution. The power and speed of digital data processing will continue to increase. The digital astronomy revolution will continue to give us more extended views of the large-scale structure of the universe. There will be no natural limit to the growth of digital surveys, until every photon coming down from the sky is separately processed and its precise direction and wavelength and polarization recorded.

So much for the impact of the digital revolution on professional astronomy. Next we turn to the amateurs. CCD cameras are now commercially available to amateur astronomers for less than a thousand dollars. For five thousand dollars you can buy a CCD system good enough to compete with the professionals. In the old days when astro-photography was done with film cameras, amateurs would travel to places remote from city lights and wait for moonless nights of exceptional clarity before attempting to take pictures of distant and dim objects. Now amateurs with CCD cameras can take good pictures of faint objects wherever they happen to be, in the light-polluted skies of New Jersey, in moonlight, or even in twilight. The decisive advantage of the CCD for amateurs is that it allows accurate subtraction of the sky background. The pictures are not yet quite as good as photographic plates taken under ideal conditions, but in average non-ideal conditions they are far better. Within a few years, as the price of the CCD comes down and the level of light pollution goes up, every serious amateur telescope will have a CCD ready to be plugged into its eyepiece. Every serious amateur will have acquired the necessary skills to use the CCD effectively.

As a result of the digital revolution, the sociology of amateur astronomy is changing. Until recently, the typical serious amateur was somebody who loved to grind mirrors, laboriously rubbing hour after hour until the optical figure was perfect. When the mirror was as perfect as it could be made, it would be put into a telescope and used for taking photographs of planets and galaxies. The object of the game was to take photographs beautiful enough to publish in *Sky and Telescope*. But now the culture of mirror grinders is dying.

Today, serious amateur astronomers are computer hackers, owning personal computers and modems and familiar with baud rates and software. They buy their mirrors ready-made. They are more at home with electronic data processing than with the fine points of mirror optics. The object of the game now is to obtain quantitative scientific data.

Because the CCD and the personal computer allow amateurs to do many things that were previously in the province of the professionals, the gap between the amateur and the professional communities has been narrowed. This is the third aspect of the digital revolution. Opportunities now lie open for professional and amateur astronomers to work together doing serious science. The professionals have many assets that the amateurs lack—big telescopes, big computers, access to government funds, and theoretical knowledge. The amateurs have two assets that the professionals lack—an abundance of instruments and an abundance of time. For any job that is labor-intensive and requires many instruments to work together, the assets of professionals and amateurs may be usefully combined. In a talk that I gave in Oxford in 1992, I suggested occultation astronomy as a promising area for collaboration between professionals and amateurs. Occultation astronomy means finding dark objects in the foreground by looking for eclipses of bright objects in the background. The universe is well suited to this mode of discovery, since it is known to be full of dark matter whose nature is one of the major mysteries of science.

Fritz Zwicky first proved the existence of dark matter in the 1930s. He had a small 18-inch telescope installed on Palomar Mountain in California for his personal use. With this little telescope he photographed the entire northern sky and compiled the first comprehensive catalogue of galaxies. He was especially interested in the galaxies that congregate in giant clusters. A typical giant cluster contains several thousand galaxies. Zwicky measured the velocities of galaxies in clusters and found that the velocities were systematically larger than they should be if the galaxies were bound to the cluster by the gravitational attraction of the visible masses. If the gravitational attraction was produced only by the visible galaxies, the clusters would fly apart. Since the clusters do not fly apart, they must be held together by some invisible mass that is five or ten times larger than the visible mass. In the sixty years since Zwicky's discovery, evidence has accumulated for dark matter pervading the universe, but nobody knows whether the dark matter consists of solid objects or cold stars or clouds of particles belonging to unknown species.

Occultation astronomy is a good technique for discovering dark matter, because we have a huge number of stars in the background and a possibly huge number of comets and planets and other opaque bodies in the foreground. Heavier dark objects can be discovered in the same way, if we look for brightening rather than darkening of background objects. Heavier objects in the foreground form gravitational lenses, bending and focusing the light from objects directly behind them. The background objects appear brightened by gravitational lensing instead of being darkened by occultation. In recent years, several groups of professional astronomers have successfully observed gravitational lensing of background stars by objects in the foreground. So far, there is no reason to believe that any of the foreground objects are not ordinary stars. To discover planets or other kinds of dark object in this way, we need more instruments and more people observing lensing events more frequently.

The way to obtain a good statistical sample of dark objects in space is to have a large number of automated telescopes watching large numbers of stars continuously for long periods of time, with datalinks to a network of computers which would correlate the telescope outputs. For such a system to be operated at reasonable cost, maintenance and supervision of the equipment by computer-wise amateurs would be extremely helpful. Professionals would have to provide overall scientific direction, central computer facilities, and data analysis. If a project along these lines were successful in finding dark objects, it would have great educational as well as scientific value. The amateur side of the enterprise could include high school science classes and small college observatories as well as individual enthusiasts and clubs. Besides the search for dark objects, many other research programs could be undertaken by an international network of automated telescopes with digital output. Any research that requires accurate monitoring of large numbers of objects could be done in a similar way. We may hope that the amateurs will bring into astronomy new ideas as well as new styles of operation. In astronomy, as well as in music and drama and all the other fine arts, it is the amateurs who sustain the culture within which the professionals can flourish.

The world of professional astronomy, the world to which the Sloan Survey belongs, is still at the beginning of the digital revolution, not yet pushing digital technology close to its limits. Our survey will use ordinary work stations to process the output at a rate of about one megabyte per second. Later surveys, using supercomputers and optical fibers to handle the output, could easily process a thousand megabytes per second. We have a long way to go before the digital revolution will have run its course. And already we see other revolutions in our future.

Until now, astronomy has traditionally been a spectator sport, with the astronomer observing celestial events but not attempting to influence the outcome. During the last few years, active intervention has ceased to be unthinkable, for two reasons. First, the public has become aware of the fact that massive biological extinctions occurred in the past and may have been associated with impacts of asteroids or comets colliding with the earth. The most famous such extinction occurred 65 million years ago, when the dinosaurs disappeared and a large impact crater appeared at the same time at Chicxulub in the Yucatán region of Mexico. Second, the collision of comet Shoemaker-Levy with Jupiter in 1994 produced wounds on that planet which were seen by millions of people either through small telescopes or on television. The public became aware of the fact that similar wounds on the earth might have made large areas of our own planet uninhabitable. The Jupiter impacts made it clear to everybody that the danger to the earth from such events is real. A rough estimate indicates that, for an average human being, the probability of being killed by an asteroid or comet impact is comparable with the probability of being killed by an earthquake. The main difference between earthquakes and impacts is that earthquakes kill small numbers of people with high probability while impacts kill large numbers of people with low probability. We take active steps to reduce the risks from earthquakes, so why not take active steps to reduce the risks from impacts?

The first step is to set up an early-warning system to detect objects that might be on a collision course with the earth. Two kinds of objects need to be detected, first the so-called Earth-crossing objects, which are either asteroids or short-period comets orbiting in the inner part of the solar system, and second the long-period comets which come from remote reservoirs in the outer part of the solar system. The search for Earth-crossing objects has already begun. It was started by Eugene and Carolyn Shoemaker, using the same little telescope on Palomar Mountain with which Zwicky discovered the existence of missing mass in clusters of galaxies fifty years earlier. If the search is given modest and steady financial support, it should be possible to obtain a complete inventory of Earth-crossing objects of substantial size within twenty years.

There are probably a few thousand such objects with diameters larger than a kilometer. It is unlikely that more than two or three of them will be as large as the Chicxulub object, which must have had a diameter of the order of ten kilometers. When the inventory is complete, we will know whether or not any of these objects is on a collision course with the earth. If we find one of them on a collision course, we shall know precisely the date of the collision and the length of the warning time within which action to avoid the collision must be taken. Since the chance that a major impact will occur within a million years is small, the warning time for the next major impact will almost certainly be longer than a hundred years. The chance that we might have a warning time as short as a year is only of the order of one in a million.

The problem of detecting long-period comets is more difficult. There can never be a complete inventory, since long-period comets appear sporadically from the depths of space, pass once through the inner solar system, and are rarely seen again. If a long-period comet is on a collision course with Earth, the warning time will be the time it takes to fall from the distance at which it is detected to Earth's distance from the sun. With our present means of detection, the warning time will generally be one or two years. To achieve a warning time of a hundred years, a far more powerful detection system will be needed. What is needed is a large wide-field telescope in orbit, a telescope that combines the wide field of the Sloan Digital Sky Survey telescope in New Mexico with the resolution and sensitivity of the Hubble Space Telescope.

The wide-field telescope in orbit does not exist and cannot be built with existing technology, but there is no reason why it should not be built with the new technologies of thin light mirrors and active optical control of mirror surfaces, technologies that are now under development and will be available within twenty years. When these technologies are perfected, it should be possible to build and launch a wide-field telescope for much less than the cost of the Hubble Telescope. A fleet of wide-field telescope satellites would allow us to carry out a new Digital Sky Survey with a hundred times the resolution of the Sloan Survey and reaching out to larger distances. As a byproduct of the new Digital Survey, we could detect long-period comets and pick out those that might endanger the earth, with warning times of the order of a hundred years. With wide-field telescopes in space to keep track of comets, and a network of smaller telescopes on the ground to keep track of Earth-crossing objects, we would have a warning time of a hundred years or longer for major impacts of both kinds. The cost of maintaining and operating such an early-warning system should not be greater than the cost of the Earth Observing System satellites that are to be deployed during the next decade to monitor the earth's biosphere and climate. It would be reasonable to spend equal amounts of money to look inward and to look outward.

What should we do when we detect a large object on a collision course with the earth? This is the question that has dominated the public discussion of the impact problem. The answer that the public has heard is to launch a barrage of nuclear missiles and deflect the object with hydrogen bombs. This is the wrong answer. Hydrogen bombs are technically unsuited to the job of deflecting an asteroid or a comet. To deflect a massive object you need to deliver momentum rather than energy. A hydrogen bomb delivers a huge amount of energy with very little momentum. To deliver momentum effectively, you need a steady gentle push, not an explosive jolt. Besides being technically ineffective, hydrogen bombs have other obvious disadvantages. The public rightly concludes that if hydrogen bombs are the answer to the impact problem, then the cure is worse than the disease. The public also suspects that the owners of hydrogen bombs are using the impact problem as a pretext for keeping the nuclear weapons industry alive.

The correct answer to the impact problem is mass-

MASS-DRIVER II

Built by Gerard O'Neill with students in the Princeton University Physics Department. 1978. Reproduced by permission of Space Studies Institute of Princeton.

drivers. A mass-driver is a machine invented by my friend Gerard O'Neill twenty years ago. O'Neill was professor of physics at Princeton University and, with the help of a class of undergraduates, built a mass-driver in the cellar of the Princeton physics department. It cost a few hundred dollars and worked beautifully. It now belongs to the Space Studies Institute in Princeton, a private research institute that O'Neill founded. It consists of a magnetic accelerator pushing small buckets along a straight track. Each bucket carries a small cargo that you wish to accelerate. At the end of the track, the cargo flies off at high speed and the empty bucket returns to the start. The weight of the cargo is adjusted to match the power and speed of the accelerator. The toy mass-driver at Princeton is only a meter long and accelerates cargo to a hundred meters per second. To deflect a comet, you would need a larger model, perhaps a hundred meters long and accelerating the cargo to one kilometer per second. The cargo would be soil or ice from the comet. The cargo would fly off into space, carrying away its momentum, and an equal and opposite recoil momentum would be delivered to the comet. The mass-driver deflects the comet by using Newton's second law of motion: to every action there is an equal and opposite reaction. The source of power would be the sun.

In order to make the system work, you would have to solve a number of engineering problems. You would have to launch the mass-driver into space and make a rendezvous with the comet. You would have to install the device on the comet, complete with solar collectors to supply it with electric power and excavators to supply it with soil. All this would take time. With a hundred years of warning, there would be enough time.

Mass-drivers are an economical solution to the impact problem because of one essential fact: The power required to deflect an object is directly proportional to the mass of the object and inversely proportional to the square of the warning time. It is the inverse-square dependence on warning time that makes the problem tractable. To give a numerical example I assume an average comet with diameter one kilometer. The mass of the comet is about a billion tons. If the output speed of the mass-driver is one kilometer per second, if the efficiency of the solar energy conversion is ten percent, and if the warning time is a hundred years, then the power required to miss the earth is ten kilowatts. Ten kilowatts is not an extravagant amount of power for a solar collector, even if it has to operate far from the sun. Ten kilowatts is power on a human scale, not on an astronomical scale. It is remarkable that ten kilowatts in the right place could save a billion lives.

Even if the impacting object was not an average comet but a monster with ten thousand times greater mass, like the object that excavated the Chicxulub crater and possibly killed the dinosaurs, the power required to deflect it would only be a hundred megawatts. A hundred-megawatt mass-driver would perhaps require an inconveniently large solar collector. For a Chicxulub-size object, which would only appear on a collision course about once in 100 million years, we might prefer to use a hundred-megawatt nuclear reactor instead of solar energy. A hundred-megawatt reactor is not a hydrogen bomb. It is small compared with present-day power reactors. If it ran for a hundred years, it would burn up only three tons of uranium fuel.

The public debate about the impact problem generally assumed that warning times would be short. With a short warning time, hydrogen bombs seemed to be the only possible response, and mass-drivers would obviously be too little and too late. But if a Chicxulub-size impacting object would actually appear with a warning time of one or two years, mass-drivers and hydrogen bombs would be equally useless. Nothing we could do would avoid the impact. Two years of warning time might allow us to save a large fraction of humanity by passive civil defense measures—stockpiling of food, medicines, fertilizers, and seedbanks to enable those who survive the immediate effects of the impact to recover and rebuild our civilization. Two years of warning time would not be useless, and might make the difference between a total collapse of civilization and a fairly rapid recovery. But a hundred years would be far better, and would allow an active defense with mass-drivers to be deployed without haste and at a reasonable cost.

The good news for humanity is that a complete system of early warning telescopes and mass-driver defenses could be built in the next century, giving us the hundred-year warning time required to solve the problem of earth-impacting objects gently and quietly. The good news for astronomy is that the early warning telescopes would not only discover impacting objects but would give us vastly extended views of the entire visible universe. Astronomy and the protection of our planet could be living together in happy symbiosis.

◈

The dominant science of the twenty-first century will be biology. Two branches of biology in particular, genetics and neurophysiology, present us with an abundance of fundamental unsolved problems that new technological tools will enable us to attack. The tools will initially be borrowed from physics and chemistry. Later in the century, new tools are likely to arise from purely biological technology. Because these problems are medically important, money will be available to people who are trying to solve them.

In genetics, one of the fundamental problems is to understand the machinery controlling the development of higher organisms. Geneticists are already making rapid progress in exploring this genetic machinery, and the biochemical architecture of the development process is understood in general terms. To understand in detail the genetic programs that guide the perfect growth of human hands and eyes will take a little longer. The detailed study of the genetic programming of human development will go hand in hand with the sequencing of human genomes. After the first human genome is sequenced, we will sequence genomes of human and other species in great variety, exploring the connections between genotype and phenotype. The building of a library of genomes for the biosphere of the earth will be an enterprise similar to the digital survey of the sky. Digital Biosphere Surveys will be as essential to the future of biology as Digital Sky Surveys will be to astronomy. In biology as in astronomy, the digital surveys will be tools for reaching a deeper understanding of our universe and its evolution. What we shall do with these tools will depend on discoveries still to be made.

The deepest unsolved problem of biology is the origin of life. When Darwin proposed his theory of evolution by natural selection in the nineteenth century, he was careful to avoid any claim that his theory could explain life's origin. In the twentieth century, various theories of the origin of life have been proposed, describing possible pathways by which populations of dead molecules might become organized into living cells. But these theories have remained vague and speculative, out of reach of experimental test. In the twenty-first century, it is likely that tools will be found to make the problem of origins accessible to experiment.

Tools from many different areas of science will be needed: from geology, new ways of finding traces of life's beginnings in the most ancient rocks; from chemistry, new ways of analyzing ancient microfossils and recovering evidence of their original composition; from microbiology, exact knowledge of the architecture of the most ancient components of modern cells, and in particular knowledge of the atomic structure of ribosomes in ancient lineages of bacteria.

The ribosome was the key biological invention, allowing genetic information to be translated into anatomical structure. The archeology of the ribosome will be a crucial testing ground for theories of early evolution. Other tools for studying life's origins will come from the exploration of the solar system. The complex chemistry of comets, asteroids, and meteorites will give detailed evidence of the chemistry of the primitive Earth out of which life arose. If we are lucky, we may find traces of ancient life on Mars. If we are very lucky, we may find indigenous life surviving in warm niches beneath the Martian surface. We might even be lucky enough to find life in the deep ocean that probably exists below the icy surface of Jupiter's moon Europa. The next century will see all these possibilities explored. The slow accumulation of facts from many disciplines will bring the origin of life out of the realm of conjecture into the realm of testable science. After all the evidence from geology, chemistry, microbiology, and planetary exploration has been brought together, it may finally be possible to simulate the origin of life with a computer and to reproduce it in a laboratory.

In neurophysiology the unsolved problems are more difficult and the way ahead is less clear. The most basic problem is to understand the principles of organization of a central nervous system. This problem cannot be attacked directly with the tools that we have now available. New tools are needed, both for collecting and for interpreting neurological data. For collecting we need new hardware borrowed from physics, to detect neural signals noninvasively with high resolution. For interpreting we need new software borrowed from computer science and mathematics, to fish meaningful signals out of a bewildering sea of noise. Neurophysiology will probably progress at a slower pace than genetics, but both disciplines are ripe for spectacular discoveries that will in turn give birth to new sciences.

When I predict that biology will be the leading science for the next hundred years, this does not mean that physics is exhausted. Physics also has its fundamental unsolved problems and is developing new tools with which to attack them. Physics is likely to progress more slowly than biology, because the tools of research in physics are more elaborate and the unsolved problems are fewer. But physics is very far from being dead. It only seems to be moving slowly because it cannot sustain the unparalleled speed with which it grew during the first half of the twentieth century. In the early twentieth century, major revolutions in physics were occurring about once every ten years. In the future development of physics, major revolutions are likely to happen as they did in the eighteenth and nineteenth centuries, about once every fifty years.

I have described the science of the twenty-first century conservatively, as a linear extrapolation of the science of today. Such a linear extrapolation is bound to fail in the long run, because the nature and fundamental objectives of science will change. If we try to imagine what science will be doing a thousand years from now, we must face the possibility that science as we know it may have ceased to exist. The thought processes of our descendants a thousand years in the future may be as alien to us as our theories of quantum mechanics and general relativity would be to Saint Thomas Aquinas. Aquinas was one of the great philosophers of his time, and our science is descended from his philosophy, but our ways of thinking have diverged so far from his in 800 years that he would find almost all our discourse unintelligible. A thousand years from now there will still be people exploring the secrets of nature in some fashion, and they may still call themselves scientists, but their tools and their purposes are likely to be so different from ours that we would hardly recognize them as colleagues following a common quest.

chapter three

TECHNOLOGY

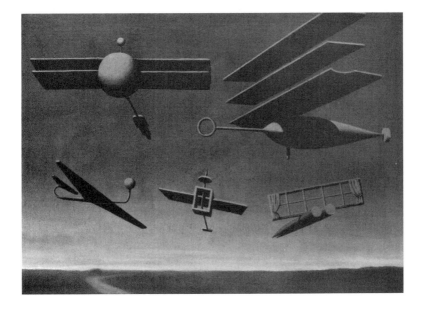

LE DRAPEAU NOIR

By René Magritte, 1937. Copyright © 1996
C. Herscovici/Artists Rights Society (ARS), New York.

TO UNDERSTAND TECHNOLOGY AS IT IS seen by people outside the technological élite, I have found science fiction more illuminating than science. Science provides the technical input for technology; science fiction shows us the human output.

I begin where Wells's *The Time Machine* left off, a hundred years ago. After Wells, the next visionary who dreamed of the future of science was the biologist J. B. S. Haldane. Haldane collected his dreams in a little book published in England in 1923 with the title *Daedalus, or Science and the Future.* The book was the text of a lecture read to the Heretics, an exclusive club of intellectuals at the University of Cambridge.

Haldane used the mythological figure of Daedalus as a symbol for the revolutionary spirit of science. Haldane was, like Wells, profoundly skeptical of the alleged human benefits of science. While Wells expressed his doubts through tragedy, Haldane expressed his through black humor. Here is Haldane explaining why Daedalus fits his theme.

It is with infinite relief that amidst a welter of heroes armed with gorgon's heads or protected by Stygian baptisms the student of Greek mythology comes across the first modern man. Beginning as a realistic sculptor (he was the first to produce statues whose feet were separated) it was natural that he should proceed to the construction of an image of Aphrodite whose limbs were activated by quicksilver. After this his interest inevitably turned to biological problems, and it is safe to say that posterity has never equalled his only recorded success in experimental genetics. Had the housing and feeding of the Minotaur been less expensive, it is probable that Daedalus would have anticipated Mendel. But Minos held that a labyrinth and an annual provision of fifty youths and fifty virgins were excessive as an endowment for research, and in order to escape from his ruthless economies, Daedalus was forced to invent the art of flying. Minos pursued him to Sicily and was slain there. Save for his valuable invention of glue, little else is known of Daedalus. But it is most significant that, although he was responsible for the death of Zeus's son Minos, he was neither smitten by a thunderbolt, chained to a rock, nor pursued by furies.

Haldane's version of the Daedalus myth departs considerably from the classical sources. We do not find in the Greek sources any support for Haldane's assertion that the birth of the Minotaur, resulting from the union of Minos's wife, Pasiphaë, with a white bull belonging to the sea god Poseidon, was contrived by Daedalus as an exercise in experimental genetics. Daedalus appears in the sources as an inventor but not as a geneticist. Nevertheless, Haldane's version of the myth is admirably suited to illustrate his thesis that the historic role of a scientist is to do the unthinkable, to overturn cherished beliefs, and to kill gods. The Minotaur is a symbol of the shocks and horrors that our well-meaning biological colleagues are about to let loose upon humanity.

My copy of *Daedalus* once belonged to Einstein. Unfortunately he left no written record of his response to it. The only evidence that he read it is a pencil-mark in the margin, presumably indicating a passage that resonated with his own thinking. It appears beside the sentence, on page 17 of the first edition, in which Haldane discusses the long-range effect of the theory of relativity on the conceptual basis of ethics: "In ethics as in physics, there are so to speak fourth and fifth dimensions that show themselves by effects which, like the perturbations of the planet Mercury, are hard to detect even in one generation, but yet perhaps in the course of ages are quite as important as the three dimensional phenomena." Einstein's theory describes space-time as a continuum with four dimensions. Haldane's mention of a fifth dimension refers to a five-dimensional version of general relativity which had been invented by Theodor Kaluza in 1921. It is the remote ancestor of the superstring theories that are fashionable seventy-five years later.

We do not know when Einstein acquired his copy of *Daedalus* or when he read it. It is unlikely that he bought it, since he was not fluent enough in English to read it for pleasure. Most probably, the book was given to him soon after it was published in 1923, either by the author or by the publisher. Haldane revered Einstein as he revered almost nobody else, expressing his reverence with the memorable words that John Reed had applied less appropriately to Leon Trotsky: "The greatest Jew since Jesus." It is consistent with what we know about Einstein's way of thinking that he believed a fundamental reform of concepts to be as necessary in ethics as in physics. If we can trust the evidence of the pencil-mark, Einstein believed this already in the early 1920s. He evidently grasped at once the main message of Haldane's book, the message that the progress of science is destined to bring enormous confusion and misery to mankind unless it is accompanied by progress in ethics. This message was unwelcome to the scientists of Haldane's time and is equally unwelcome today.

Haldane did not know much about theoretical physics, but he had expert first-hand knowledge of warfare and physiology, the two subjects that provide the core of his argument. His knowledge of war and physiology was not theoretical but practical, gained by risking his life repeatedly in the trenches of France and the coal mines of Staffordshire. He began his education in physiology by serving as his father's assistant in investigations of the effects of noxious gases to which miners and sailors were exposed in coal mines and submarines. Later, he took part in similar investigations of the effects of lethal gases on soldiers in the first world war. He fought with legendary bravery as an infantry officer on the Western Front. When he writes in *Daedalus*, "Death has receded so far into the background of our normal thoughts that when we came into somewhat close contact with it during the war most of us failed completely to take it seriously," we can be sure that he is recording his personal experience.

Daedalus begins with an artillery bombardment on the Western Front, the shell bursts nonchalantly annihilating the human protagonists who are supposed to be in charge of the battle. This opening scene epitomizes Haldane's hard-headed view of war. And likewise at the end, when the biologist in his laboratory, "just a poor little scrubby underpaid man groping blindly amid the mazes of the ultramicroscopic," is transfigured into the mythical figure of Daedalus, "conscious of his ghastly mission and proud of it," this closing scene epitomizes Haldane's hard-headed view of science. Haldane was saying that the destiny of the scientist is to turn good into evil, that the horrors of the first world war are not an isolated phenomenon but only an example of the disruptive consequences that we may constantly expect to emerge from the progress of science. Now, seventy-four years later, we are beginning to see more clearly what he had in mind.

The decision of the United States Congress to kill the Superconducting Supercollider project came as a shock to many of my scientist colleagues, but it would not have surprised Haldane. At the time when Haldane was writing in the 1920s, science was intensely unpopular in England. Science was identified in the public mind with the technological carnage of the recent war. The first world war was seen as peculiarly evil, because

LA GUERRE MIASMATIQUE

*By Albert Robida, 1900. From La vie électrique. Reproduced by
permission of the Houghton Library, Harvard University.*

the organizers and promoters of the slaughter were old men while the victims were young. The public blamed scientists in general, and chemists in particular, for the invention of explosives and poison gases which killed or scarred a whole generation of young Englishmen. Scientists were seen as a privileged priesthood, callously profiting from the misery of the unprivileged. Forty years later in America, a similar hatred of science was aroused in the generation of young people who experienced the consequences of technology in the Vietnam war and felt themselves to be victims.

Today, science has once again turned good into evil. This time the evil is not a war, but a civilian technology that systematically widens the gulf between rich and poor, deprives uneducated young people of jobs, and leaves large numbers of young mothers and children homeless and hopeless. The evil is to be seen in many places around the world, especially in the great cities of North and South America. When one walks through the streets of New York after dark during the Christmas season, one sees the widening gulf at its starkest. The brightly lit shop windows are filled with high-tech electronic toys for the children of the rich, and a few yards away, the dark corners of subway entrances are filled with the dim outlines of derelict human beings that the new technology has left behind. In every large American city, such contrasts have become a part of everyday life.

When I first arrived in America fifty years ago, rich and poor people were less estranged and less afraid of one another, the feeling of belonging to a community was stronger, the rich had fewer locks on their doors, and the poor had roofs over their heads. Since those days, wealth has accumulated and society has decayed. It is as Haldane said, "The tendency of applied science is to magnify injustices until they become too intolerable to be borne, and the average man, whom all the prophets and poets could not move, turns at last and extinguishes the evil at its source."

My scientist friends may justly protest that the calamities of American society are caused by drugs, or by guns, or by racial intolerance, or by illiteracy, or by bad schools, or by broken families, rather than by science. It is true that the immediate causes of social disintegration are moral and economic rather than technical. But science must bear a larger share of responsibility for these evils than the majority of scientists are willing to admit. When we look at historical processes on a time-scale of fifty or a hundred years, science is the most powerful driving force of change. Because of science, machines have displaced unskilled manual workers, and computers have displaced unskilled clerical workers, in all branches of industry and commerce. Because of science, the traditionally conservative middle class of well-paid blue-collar industrial workers has almost ceased to exist. Because of science, jobs paying enough to support a family in comfort are no longer available to young people without higher education, unless they happen to be gifted with special talent as baseball players or rock stars. Because of science, families with access to computers and to higher education are rapidly becoming a hereditary caste, the children inheriting these advantages from their parents. Because of science, children deprived of legitimate opportunities to earn a living have strong economic incentives to join gangs and become criminals.

In Trenton, only a few miles away from the academic oasis of Princeton where I live, many of the children of the inner city seize their first chance of financial independence at the age of nine, when they are recruited by drug dealers to serve as scouts to give warning of police raids. The fateful turning point in their lives comes in the summer between third and fourth grades, when the gang displaces the school as their main source of education. Technological change, driven by science, has been the primary cause of these revolutions in the economic base of society. After technological change has closed down industries and destroyed jobs, the decline of morality and the erosion of discipline follow as secondary causes of social breakdown.

⚜

Haldane did not foresee the computer, the most potent agent of social change during the last fifty years. He expected his Daedalus, destroyer of gods and of men, to be a biologist. Instead, the Daedalus of this century turned out to be John von Neumann, the mathematician who consciously pushed mankind into the era of computers. Von Neumann knew well what he was doing. Soon after the end of the second world war, he started the Princeton computer project. Like Haldane's Daedalus, he had dreams that went far beyond the scientific instrument that he was building in Princeton. He spoke and wrote much about automata. His automata were abstract generalizations of a computer. An automaton was a machine that could not merely compute but carry out actions in the real world as instructed by its program. Von Neumann saw that there was no limit to the scale and complexity of actions that automata could perform. His computer was only a small step toward the realization of his dream of automata guided by artificial intelligence.

Beyond the intelligent automaton was another dream, the self-reproducing automaton. Von Neumann proved with mathematical rigor that a self-reproducing automaton was possible, and enunciated the abstract principles that would govern its design. He dreamed that the creation of self-reproducing automata would be a boon to mankind, abolishing hunger and poverty all over the earth, providing us with obedient slaves to satisfy our needs. Self-reproducing automata could build our homes, cook our food, and wait on us at table. But Von Neumann, like Haldane's Daedalus, was destined to turn good into evil. An unfriendly critic might say that the hidden purpose of Von Neumann's dream of self-reproducing automata was to make all humans superfluous except for mathematicians like himself. In the end, even the mathematicians, who would initially be needed in order to design the automata, might also turn out to be superfluous.

Two developments that Von Neumann did not foresee were the personal computer and the computer-game software industry. These two side-effects of his activities have grown with explosive speed. Like other rapid technological changes, they have brought with them both good and evil. On the good side, they have given us computers with a human face, computers accessible to ordinary people for profit or for fun. Von Neumann never imagined that computers could be humanized to such an extent that mothers would use them to print birth announcements and schoolchildren would use them to do homework. On the evil side, the home-computer industry has widened the gap between rich and poor. The child of computer-owning parents grows up computer-literate and is showered with opportunities to enter the world of high-tech education and industry. The child without access to a home computer is left behind. Computer illiteracy is an additional barrier that a poor child has to overcome in order to earn an honest living.

Haldane looked unflinchingly at the evil consequences of science, both in war and in peace, but he was not on that account a pessimist. He did not believe in gloom and doom. The final message of *Daedalus* is not gloomy. Haldane had far too much respect for ordinary people to be a pessimist. He admired the toughness of the ordinary soldiers who fought under his command in France. He loved India, and chose to live there at the end of his life, because he had a natural affinity for people who remained cheerful in spite of hardships. When he was in hospital recovering from an unsuccessful cancer operation, he wrote a cheerful poem with the title "Cancer's a Funny Thing." His innate optimism shines through the superficial cynicism of *Daedalus*. He considered the consequences of science to be mostly evil and dangerous, but, because science forces us to overcome these evils and dangers, he said, "Science holds in her hands one at least of the keys to the thorny and arduous path of moral progress." The final message of *Daedalus* is that ordinary people can turn evil into good if they have the necessary courage and moral leadership. Haldane had no doubt that they have the courage, and he intended to give them the moral leadership.

❧

Haldane was surely right in expecting that the most profound shocks to human society would come from biology. He mentioned two biological shocks in particular, the genetic engineering of microbes that would invade the oceans and replace agriculture as a source of food, and the technology of ectogenesis that would replace motherhood as a source of babies. Neither of these shocks has yet happened. Although the technologies of genetic engineering and ectogenesis are developing rapidly and are already in use to a limited extent, giving us new drugs and in vitro fertilization, agriculture and motherhood are still alive and well. It does not now seem likely that they will be displaced by biotechnology within the next century.

Still, there can be little doubt that genetic engineering and ectogenesis are destined to give us rude jolts in one way or another. Sooner or later, genetic engineering will allow us to create new species of plants and animals according to our whim, and to choose the genetic endowment of our children. Ectogenesis will allow us for the first time to achieve full equality of biological status between men and women. Either one of these innovations will bring more profound changes to human society than the advent of personal computers. Both innovations are likely to sharpen the social conflicts between liberal and conservative, between believer and unbeliever, between rich and poor. Both innovations may be resisted successfully for a while by legal restraints, by religious prohibitions, or by physical violence. And in the long run, both innovations are likely to prevail over the opposing forces, at least in some places and in some segments of society. Nobody can predict how long the run will be before these things happen. Haldane expected them to happen before the end of the twentieth century. He turned out to be wrong. I am probably erring on the side of caution when I say that they will happen before the end of the twenty-first.

The awesome power that genetic engineering will one day place in our hands was foreshadowed recently by some experimenters at the University of Basel in Switzerland. Walter Gehring and his students were studying the effects of the *eyeless* gene in fruit flies. The gene is called *eyeless* because its absence causes flies to grow without eyes. The gene actually causes eyes to

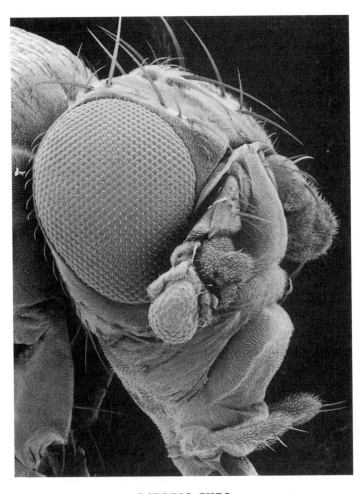

ECTOPIC EYES

*On this head of a fruit fly, eye facets are growing on the antenna
(to the lower right of the eye). 1995. Reproduced by permission of
W. J. Gehring, G. Alder, and P. Callaerts, Biozentrum,
University of Basel, Switzerland.*

grow. Gehring and the students inserted the gene into various tissues of embryonic flies, and the embryos grew into flies with eyes sprouting from their legs or wings or antennas. The superfluous eyes were anatomically perfect. This gene somehow commands a whole army of several thousand subordinate genes to perform the miracle of creating an eye.

But this was only the first miracle. Gehring's group had previously discovered that the *eyeless* gene in fruit flies is chemically similar to a gene *small-eye* in mice and to a gene *aniridia* in humans. The absence of *small-eye* causes mice to grow with defective eyes, and the absence of *aniridia* causes blindness in human babies. The second miracle occurred when Gehring put the *small-eye* mouse gene into the leg of a fruit-fly embryo. The embryo grew into a fly with a perfect fruit-fly eye attached to its leg. So the identical *small-eye* gene is able to command the creation of a mouse eye in a mouse and the creation of a fruit-fly eye in a fruit fly. This shows an amazing flexibility in the power exercised by the gene. The basic design of a mouse eye is totally different from the design of a fruit-fly eye. Every detail in the anatomy of the two eyes is different. The mouse eye is like a human eye, with a retina and a lens, the lens focusing an optical image onto the retina. The fruit-fly eye is an array of hundreds of independent units, each detecting light from a particular direction but without any sharp focus. The experiment shows that the *small-eye* gene is not concerned with anatomical structure but with abstract function. The gene commands "Create an eye" in some abstract language that the mouse genome can translate into mouse anatomy and the fruit-fly genome can translate into fruit-fly anatomy. The language of the genes is demonstrated to be far more abstract and flexible than anybody had imagined before these experiments were done.

Another step forward in the understanding of the genetic language is described in a paper by three California experts, Eric Davidson, Kevin Peterson, and Andrew Cameron, discussing the early evolution of animals with well-defined anatomical structure. They examine the sixteen major groups (phyla) of many-celled organisms and observe that twelve of the sixteen phyla use a peculiar method of growing adult creatures from larval embryos. The peculiar method is called "indirect development," meaning that the adult phase has no structural resemblance to the larval phase out of which it grows. The two phyla that are most familiar to us, the vertebrates (animals with backbones) and the arthropods (insects, crustaceans, and spiders) are exceptions to the general rule. The metamorphosis of tadpoles into frogs and of caterpillars into butterflies is direct, the adult retaining through the metamorphosis the basic body plan of the larva.

The twelve phyla that develop indirectly are more primitive and probably closer to the root of the evolutionary tree than the two familiar exceptions. In an indirectly developing creature, the larval form is small and simple, and contains a package of undifferentiated cells set aside for later use. After the growth of the larva is completed, a genetic switch turns on the development of the set-aside cells, which then differentiate and grow into the adult, following a blue-print unrelated to the larval anatomy. The function of the larva is merely to provide a life-support system to the adult during the vulnerable phases of its early growth.

After describing these facts concerning the embryonic development of modern organisms, the three California geneticists go on to extrapolate the facts of embryonic development into a hypothesis of ancient evolution. I call their hypothesis the two-stage theory of multicellular evolution. The two-stage theory says that the evolution of many-celled creatures occurred in two jumps. The first jump was the evolution of a creature that later became the larval form of indirectly developing animals. The second jump was the evolution of a separate genetic architecture for the development of adult forms. The first jump uses a hard-wired pattern of genetic control, with batteries of genes switched on in sequence to direct the division and differentiation of cells all the way from the single fertilized egg to the complete larva. The larval forms are small, usually containing only a few thousand cells. The hard-wired pattern of control probably did not work well beyond this limit. As a result, according to the theory, further evolution had to wait until the second jump, when a new and more abstract genetic language was invented to direct the organization of the larger and more elaborate structures that are seen in adult forms. The more abstract language contains genes like *eyeless* and *small-eye* that are capable of organizing the diverse structures of eyes in species as different as mice and fruit flies.

The genetic language of the second jump is like the language of software in a computer. The step from the first to the second jump was like the step from computers using hard-wired instructions to computers using software. The function of the larval stage in the evolution of indirectly developing creatures was to provide life-support to the cells that carried out the organization of the adult stage, so that the genetic apparatus of the adult stage could evolve through rapid and drastic changes of program without killing the organism. The Cambrian explosion—the sudden appearance in the

Cambrian era 540 million years ago of the entire diversity of many-celled phyla—was the result of the successful completion of the second jump. Once the abstract genetic language had been perfected, a marvelous menagerie of alternative adult body plans could be programmed and evolve by natural selection within a few million years. After the Cambrian epoch, no radically new anatomical patterns have appeared. In arthropods and vertebrates, the primitive pattern of indirect development was probably replaced by direct development at some later period, after the basic adult body plans had been fixed.

The two-stage theory describes the way evolution works in many areas of science and technology. The theory applies to the evolution of computers and software in the twentieth century as well as to biological evolution in the Precambrian era. Another striking analogy may be seen between the two-stage evolution of many-celled organisms in the late Precambrian and the two-stage evolution of single-celled organisms a billion years earlier, when eukaryotic cells with nuclei and elaborate internal structure arose from the smaller and simpler prokaryotic cells by a process of cellular fusion. Lynn Margulis has studied the origin of eukaryotic cells in detail and has argued convincingly that eukaryotes arose from a two-jump process, beginning with parasitic invasion of one prokaryote by another and ending with complete symbiosis. A billion years before that, the origin of life may also have been a two-jump process, with metabolism as the first jump and replication as the second. The first jump would have involved enzymes and protein molecules, the second jump genes and nucleic acids. But I resist the temptation to digress further into two-stage models of evolution.

The Basel experiments gave us only a first glimpse of the abstract language of the genes. We do not yet have any idea of the syntax and grammar of the language, or of the limits to its powers of expression. Many years of laborious exploration lie ahead, before we shall be able to read the language and understand its nuances. But the time will come, probably within a few decades, when we shall read it fluently. The tools for reading and writing DNA are analyzers and synthesizers which already exist in every laboratory where molecular biology is studied. As soon as we are able to read the language, we shall also be able to write it. The ability to write the DNA language with a full understanding of its subtleties will mean the ability to play God, to build a Jurassic Park and fill it with dinosaurs of our own design.

We must expect that, as soon as genetic engineering

at the creative level becomes practical, it will move from the laboratory and the hospital into the entertainment industry. Toys and games are the fastest way to make any new technology popular and profitable. Computer technology did not grow explosively until it was incorporated into games. Genetic engineering will not grow explosively until it moves into toyshops and theme parks. Jurassic Park may be a fantasy, but the universal popularity of dinosaurs is real. Little genetically engineered dinosaurs may be as ubiquitous in the lives of our great-grandchildren as little plastic dinosaurs are in the lives of our children. Animal-rights activists may fight against the private ownership of dinosaurs, but it will be difficult to argue that giving a child a dinosaur is more cruel than giving a child a puppy.

Fifty years ago, the philosopher Olaf Stapledon published a novel, *Sirius*, which explores some of the depths of loneliness and alienation to which genetic engineering might lead. Stapledon knew nothing of DNA and molecular biology, but he foresaw the possibility of genetic engineering and saw that it would give rise to severe dilemmas. His hero, Sirius, is a dog endowed with a brain of human capacity by doses of nerve-growth hormone given to him in utero. His creator raised him as a member of his own family together

BOY WITH TRICERATOPS

Illustration by Louis Darling, from The Enormous Egg by Oliver Butterworth, 1956. Reproduced by permission of Little, Brown & Company.

with a human daughter. Sirius retains the instincts and loyalties of a dog but thinks like a human. His foster-sister understands him and communicates with him on the human level, but humans who do not know him are frightened and see him as a fearsome monster. His story is a mixture of triumph and tragedy. The triumph is the psychological insight that he achieves by harmonizing the two sides of his nature into a seamless whole. He understands humans more deeply than any mere human can understand them, and learns to accept his own predicament with philosophical detachment. The tragedy is the impossibility of finding a place for him in the world of ordinary humans and ordinary dogs. Sirius can neither be fully human nor fully dog. The tragedy is played out against a background of austere lives in wartime Wales. When his foster-sister is drafted for war service, he is left in solitude among strangers, and his life ends in grief and violence.

At present the two new technologies that are turning the world upside down, the technology of computers launched by von Neumann and the technology of genetic engineering launched by Crick and Watson, are running along separate paths. Computer technology promises to give us self-reproducing automata, machines built out of metals and semiconductors that could replace our existing machines and satisfy our needs more cheaply and flexibly. One type of self-reproducing automaton might be an intelligent house-builder that creates a custom-built house following the wishes of the customer. Another might be an intelligent automobile that takes us where we wish to go while safely avoiding collisions and relieving us from the tedium of driving. A third type might be a solar-energy collector that makes electricity and feeds it into the local power grid. Underlying all these promises is the belief that a machine that reproduces itself out of commonly available materials will ultimately be cheap. If the technology of self-reproducing machines is cheap, it can be a liberating force for poor people and poor countries all over the earth. If the technology is expensive, it will only be another toy for the rich.

Meanwhile, the technology of genetic engineering is making similar promises. A genetically engineered bacterium or fungus designed for a particular purpose is in essence nothing more than a self-reproducing automaton made of protein and nucleic acid instead of metal and silicon. Genetically engineered automata are likely to be particularly effective in the business of chemical processing. They could be programmed to metabolize unwanted chemicals polluting land or water or air. They could convert pollutants efficiently into harmless or useful byproducts. Genetically engineered scavengers could replace existing chemical and garbage-disposal industries at the same time as self-reproducing machines replace existing construction and transportation industries. The technologies of automata and of organisms are engaged in a competition to take the leading role in the industrial revolution of the twenty-first century. Until now, computers and automata have been in the lead, but molecular biology and genetics are not far behind.

Neither genetics nor computers are likely to win the

race outright. As the physical structures at the heart of modern computing become smaller, while the chemical structures at the heart of genetic engineering become more versatile, the two technologies are beginning to overlap and to merge. It is likely that the winning designs for an intelligent solar-energy machine or an intelligent garbage-disposal machine will make use of electronic and biological tools working together. The self-reproducing machine will be partly made of genes and enzymes, while the genetically engineered brain and muscle will be partly made of integrated circuits and electric motors. In the end, physical and biological components will be so intimately entangled that we will be unable to say where one begins and the other ends. Sometime before the end of the twenty-first century, the industrial revolution based on the symbiosis of metal and silicon with nerve and muscle will begin. If all goes well, this revolution will bring beauty to industrial landscapes and wealth to cities all over the earth. And if all does not go well, as usually happens in human affairs, then beauty and wealth will continue as before to be distributed unequally.

Nine years after *Daedalus*, Haldane's friend Aldous Huxley published *Brave New World*. Huxley translated Haldane's ironic essay into a best-selling novel and added some new ideas of his own. Huxley's classic

HUMAN NERVE CELLS GROWING ON
AN INTEGRATED CIRCUIT

*This scanning electron micrograph was taken in the
course of research into possible electronic devices of the
future that would combine organic and inorganic
components. 1985. Reproduced by permission of
Synapter/Science Photo Library.*

description of a biotechnological utopia borrowed from Haldane the technologies of genetic engineering and ectogenesis. The new features that Huxley added were the cloning of large numbers of identical human beings, the free supply of euphoric drugs with no deleterious side-effects, and the benevolent tyranny of a world government dedicated to the slogan "Community, Identity, Stability."

The story begins with the Assistant Director of Hatcheries escorting a group of young technicians on a tour of the apparatus for cloning and hatching human embryos. The production lines produce various grades of embryo, running all the way from alpha-plus to epsilon-minus, their physical and mental capacities specialized for the various roles that they are to perform in society. This idyllic scene of socialist bureaucracy is supposed to lie six hundred years ahead of us. Six hundred years is plenty of time for the social problems of our own era to be solved, for the history of divisive struggles between nations and races to be forgotten, and for the spark of human individuality to be extinguished. Six hundred years from now, we shall have achieved the age-old dreams of perpetual peace and the greatest happiness of the greatest number. As the price of this achievement, we shall make some small sacrifices. We shall sacrifice some of the human qualities that might disturb the stability of our society. We shall sacrifice intellectual curiosity and political discontent. We shall sacrifice the personal ambitions that cause us to fight and quarrel and stir up revolutions. We shall sacrifice the three ungovernable passions that brought us so much grief in the past: science, art, and religion.

The hero of *Brave New World* is John, a young man who grew up on an Indian reservation in New Mexico. The reservation is inhabited by primitive peoples and maintained by the benevolent world government as a tourist attraction. It exists so that the civilized tourists can observe from a distance the nasty and brutish lives of people who have the misfortune to be unprotected by the cushions and comforts of technology. On the reservation, traditional religions and traditional customs are tolerated. John was born without the aid of ectogenesis, from a human mother. He found on the reservation an ancient volume containing the plays of Shakespeare, and acquired from it a taste for poetry and a tragic view of life. The theme of the novel is John's encounter with the civilized world. He makes friends in the civilized world and attempts to convert them to his way of thinking. He is asking them to rebel against everything they have been taught, to claim their personal freedom by cutting themselves off from their society, to choose pain and loneliness as the price of human dignity. Inevitably he fails. Not one of his friends understands him. For his friends, human dignity and tragedy have no meaning. For them, pain and loneliness are not tragic but merely absurd. So they laugh at his old-fashioned ideas and leave him in the final scene alone with his dignity, swinging in the wind at the end of a rope.

Brave New World gives us a dramatic view of a future in which the technology made possible by science brings science to a halt. This future is consistent with the more remote future seen by the Time Traveler in Wells's *Time Machine.* After the disruptive influence of science has been permanently tamed by the triumph of bureaucracy and eugenics, it is easy to imagine human society remaining stuck in the rigidly conservative caste system of *Brave New World* for thousands of centuries, until the slow processes of mutation and degeneration reduce our species to the condition of the Eloi and Morlocks as the Time Traveler encounters them in the year Anno Domini 802701.

The fantasies of Wells and Huxley were based on the same idea, that a species adapting itself too perfectly to a static ecological niche is doomed to stagnation and ultimate extinction. Their nightmares describe a possible future for our species, if we succeed in building around ourselves a protective cocoon that shields us from the winds of change while our mental faculties dwindle. A future of senile dementia is as possible for the species as it is for the individual.

And yet, when I compare these visions of a static and immobile humanity with the actual turbulence of human history, I am tempted to exclaim with Winston Churchill, "What kind of people do they think we are?" Churchill was addressing this remark to the people of England in 1940, when Hitler was generously inviting us to make peace with him after he had conquered France. The same remark applies equally well to the human species as a whole, when learned experts pronounce us doomed to a future of stagnation or impoverishment. The human species has a deeply ingrained tendency to prove the experts wrong. Only ten years ago, the experts were proclaiming the Soviet Union to be a stable and conservative society. Now, ten years later, for better or for worse, the Soviet Union is gone with the wind and the experts are still trying to explain how it could have happened.

So far as I know, the only writer who correctly predicted what happened to the societies of Eastern Europe was not a professional expert but a novelist. His name was Bruce Chatwin and he published his prediction in the year 1988 in his novel *Utz*. Here is Chatwin's story of what was to happen in Prague a year and a half after his book was published. His hero Utz lived in Prague and knew something of the history of Central Europe.

I think it was Utz who first convinced me that history is always our guide for the future, and always full of capricious surprises. The future itself is a dead land because it does not yet exist. When a Czech writer wishes to comment on the plight of his country, one way open to him is to use the fifteenth-century Hussite Rebellion as a metaphor. I found in Prague Museum this text describing the Hussites' defeat of the German Knights: "At midnight, all of a sudden, frightened shouting was heard in the very centre of the large forces of Edom who had put up their tents along three miles near the town of Žatec in Bohemia, in the distance of ten miles from Cheb. And all of them fled from the sword, driven out by the voice of falling leaves only, and not pursued by any man". As I scribbled this in my notebook, I seemed to hear again Utz's nasal whisper: "They listen, listen, listen to everything but . . . they hear nothing". He had, as usual, been right. Tyranny sets up its own echo-chamber, a void where confused signals buzz about at random, where a mur-

mur or innuendo causes panic. So, in the end, the machinery of repression is more likely to vanish, not with war or revolution, but with a puff, or the voice of falling leaves.

At the time when Chatwin was hearing the voice of falling leaves in the forest of Žatec, the professional experts in East and West were still listening to everything and hearing nothing. Only two years later, the voice of falling leaves was heard by the rulers of totalitarian regimes, not only in Prague but also in Warsaw and Berlin and Moscow. The moral of Utz's story is that the best way to predict the future of human society is to study the past. The past shows us that humans are an ungovernable crowd, averse to logic and discipline. It is unlikely that any totalitarian world government of the future, or any dogmatically imposed system of beliefs, will last longer than the theocratic utopias of the past. Theocratic utopias have usually lasted about twenty years, just long enough for a new generation of children to grow up and rebel against the power of the founding fathers. The Soviet Union lasted for seventy years, but its utopian ideology was already waning after the first twenty.

Saul Bellow, in his speech replying to the award of

the Nobel Prize for Literature in 1976, spoke of the failure of modern literature to give credit to ordinary people for their resistance to the homogenizing effects of modern technology.

> We do not, we writers, represent mankind adequately. We do not think well of ourselves. We do not think amply about what we are. Essay after essay, book after book, maintain the usual thing about mass society, dehumanization, and the rest. How weary we are of them. How poorly they represent us. The pictures they offer no more resemble us than we resemble the reconstructed reptiles and other monsters in a museum of paleontology. We are much more limber, versatile, better articulated; there is much more to us; we all feel it.

The human children who defeat the reconstructed reptiles and evil scientists in the best-selling science-fiction soap opera *Jurassic Park* are teaching us the same lesson as Saul Bellow. *Jurassic Park* is like *Brave New World,* another novel creating a myth about future disasters arising from the practice of genetic engineering. But *Jurassic Park* is a cheerful adventure story while *Brave New World* is a tragedy. Even if *Brave New World* is a greater work of literature, *Jurassic Park* comes closer to being a true statement of the human condi-

FROM JURASSIC PARK

*Photograph by Murray Close, 1992. Copyright © 1992 by
Universal City Studios, Inc. Reproduced by permission of MCA
Publishing Rights, a division of MCA, Inc.*

tion. As every parent and grandparent knows, human children are born rebels.

The next revolutionary technology that might come after genetic engineering is neurotechnology, the development of tools for exploring and manipulating the human brain. Neurotechnology already exists, but its methods are crude and its powers limited. In the future it will grow less crude and more powerful. It may or may not give birth to a technology of enormous power which I call radiotelepathy. Radiotelepathy may turn out to be impossible. I discuss it here as an example of the many possible revolutions to which neurotechnology might lead. Radiotelepathy, if it is possible, could grow out of the science of neurophysiology just as genetic engineering grew out of molecular biology. After the principles of genetics had been explored and understood, the way was open to develop the technology of gene splicing and to use it for practical purposes. In the same way, after the organization of the central nervous system has been explored and understood, the way will be open to develop and use the technology of electromagnetic brain signals.

The idea of radiotelepathy first appeared, so far as I know, in the science-fiction novel *Last and First Men,* written by Olaf Stapledon in 1931, one year before the publication of *Brave New World* and thirteen years before *Sirius.* Stapledon imagined a form of life evolving on the planet Mars, in which the cells of a multicellular creature communicate with each other by means of electric and magnetic fields instead of by physical contact and chemical exchange. A Martian is a green cloud composed of tiny droplets which Stapledon calls "sub-vital units." The sub-vital units transmit and receive electromagnetic fields. The fields perform the functions of muscles and nerves, so that the cloud behaves as a coherent individual. The same idea was used later by the astronomer Fred Hoyle in his science-fiction story, "The Black Cloud." But Stapledon pursued the idea further. In Stapledon's imaginary history, Martians and humans live side by side on the earth in a state of intermittent warfare for thousands of years, until both species are almost destroyed. And then, after a long period of recovery and readjustment, a new human species appears with Martian sub-vital units incorporated as symbiotic organelles into its brain cells. The new human species is able to communicate from one brain to another by radio. It has acquired the Martian radiotelepathy sense organ in an otherwise human body.

It is remarkable that Stapledon imagined the Martian units invading human cells as parasites and evolving into symbiotic partners, before the experts in evolutionary biology discovered that the mitochondria in human cells have actually evolved from alien invaders to symbionts in precisely this manner, one or two billion years earlier. The biologists of Stapledon's time knew nothing of the molecular evidence that would demonstrate the symbiotic origin of mitochondria.

Stapledon's story of the telepathic Martians is only a myth. But Stapledon's myth contains features that might one day come true, if the science of neurophysiology advances as rapidly in the future as the science of molecular biology has advanced in the last fifty years. The chief barrier to progress in neurophysiology is the lack of observational tools. To understand in depth what is going on in a brain, we need tools that can fit inside or between the neurons and transmit reports of neural events to receivers outside. We need observing instruments that are local, nondestructive, and noninvasive, with rapid response, high band-width and high spatial resolution. We need to invent the terrestrial equivalent of a Martian subvital unit.

There is no law of physics that declares such an observational tool to be impossible. We know that high-frequency electromagnetic signals can be propagated through brain tissue for distances of the order of centimeters. We know that microscopic generators and receivers of electromagnetic radiation are possible. We know that modern digital data-handling technology is capable of recording and analyzing the signals emerging from millions of tiny transmitters simultaneously. All that is lacking in order to transform these possibilities into an effective observational tool is the neurological equivalent of integrated-circuit technology. We need a technology that allows us to build and deploy large arrays of small transmitters inside a living brain, just as integrated-circuit technology allows us to build large arrays of small transistors on a chip of silicon.

I give the name radioneurology to this hypothetical future technology of observing neural processes inside a brain by means of locally deployed radio transmitters. Radioneurology is in principle only an extension of the existing technology of magnetic resonance imaging, which also uses radio-frequency magnetic fields to observe neural structures. Radioneurology would allow observation with far higher resolution in both space and time.

A rough estimate based on the available band-width indicates that a million transmitters could be monitored through each patch of brain surface with size equal to the radio wave-length. The factor of a million is the ratio between the radio band-width, of the order of hundreds of millions of cycles per second, and the band-width of a neuron, of the order of hundreds of cycles. It is possible that the apparatus of radioneurology could be developed as integrated circuits were developed, by steady refinement of physical and chemical techniques. On the other hand, radioneurology might take advantage of electric and magnetic organs that already exist in many species of eels, fish, birds, and magnetotactic bacteria. In order to implant an array of tiny transmitters into a brain, genetic engineering of existing biological structures might be an easier route than microsurgery.

Radioneurology is not the same thing as radiotelepathy. Radioneurology is supposed to be a scientific tool, helping us to understand the architecture of central nervous systems. It translates information about neural processes into electromagnetic signals that can be received and decoded outside the brain. It is a technology for the passive observation of mental processes. However, once the technology of passive observation exists, it will easily be extended to active intervention. When we know how to put into a brain transmitters translating neural processes into radio signals, we shall also know how to insert receivers translating radio signals back into neural processes. Radiotelepathy is the technology of transferring information directly from brain to brain using radio transmitters and receivers in combination. Radioneurology is a comparatively harmless technique, useful for scientific research and for medical diagnosis, but radiotelepathy will raise acute questions of ethics.

Stapledon imagined radiotelepathy both as a force for evil and a force for good. In the Martian species where it originated, it was a force for evil, dragooning the Martian population into a totalitarian war machine dedicated to the destruction of humanity. After the final defeat of the Martians, when radiotelepathy spreads to the human species, it becomes a force for good, allowing humans to understand one another better and settle their differences peacefully. In the real future, if radiotelepathy turns out to be possible, it will be a powerful force that can be turned either to good or to evil purposes. Like genetic engineering, radiotelepathy will be controversial even when it is applied to nonhuman species, and will cause severe problems when it is applied to humans.

Laws regulating radiotelepathy must begin with a guarantee of privacy. Every person capable of radiotelepathy must be provided with reliable means for switching off both transmitters and receivers. This guarantee would go some way toward protecting the individual from telepathic eavesdropping and coercion. But it would not go far enough. Just as with the application of genetic engineering to humans, the application of radiotelepathy would raise the thorniest questions when it involves children. If, as is likely, the ability to master the radiotelepathic language has to be learned in early childhood like the mastery of spoken language, then the legal principle of informed consent has no meaning. Babies cannot give informed consent to their own birth and up-bringing. Only after they are grown up can they look back and decide whether they are lucky pioneers of a new world or unlucky victims of their parents' ambition. Big jumps in technology or in evolution always have costs that are incalculable.

chapter four

EVOLUTION

SOMETHING FROM SLEEP II
*By David Wojnarowicz, 1988. Reproduced by
permission of the artist.*

All the world's a stage,
And all the men and women merely players.
They have their exits and their entrances.
And one man in his time plays many parts,
His acts being seven ages.

THE THEME OF THIS CHAPTER COMES from a passage in Shakespeare's *As You Like It.* Shakespeare's seven ages of man are infant, schoolboy, lover, soldier, justice, lean and slippered pantaloon, and second childhood. I am playing a role half-way between the fifth and sixth ages, an old scientist pretending to be a sage. If I were Shakespeare I would already have been dead for seventeen years. Shakespeare's picture of us, men and women, strutting briefly on the stage of history and playing parts that we do not fully understand, is a good summary of the human condition. I use the framework provided by Shakespeare to introduce my view of the evolution of humanity.

The central fact of our situation is that we are living on many different time-scales. Humanity has survived in the past and must learn to survive in the future by dealing with problems on many time-scales simultaneously. My seven ages of man are not the seven parts of an individual life but the seven different time-scales on which our species has adapted to the demands of nature. Each time-scale makes different demands, and the demands of the various time-scales are often in conflict with one another. This is the main reason for the complexity of human nature. We are the only species that is conscious of the passage of time and of our own mortality, the only species with an awareness of the future.

I choose my seven ages of man arbitrarily to be ten years, a hundred years, a thousand years, ten thousand years, a hundred thousand years, a million years, and infinity. I choose these particular periods only because we are accustomed to measure historical time in decades and centuries and millennia. The fact that we use a decimal system to measure time is not essential. The essential point is that our history is dominated by different processes at each of the different time-scales. When we think about our future, we must first understand that the future, like the past, comes with all these different time-scales. It is important, in any discussion of the future, to distinguish carefully the time-scales on which various things may happen. Only after we have specified the time-scale can we speak of future possibilities with some degree of clarity.

I chose the shortest time-scale to be ten years, the same length as Shakespeare's seven periods in the life of an individual. This is a natural time-scale for human activities. Anything shorter belongs to the present rather than to the past or the future. I chose the second-longest time-scale to be a million years because that is roughly the age of our species. The past and future extend much further than a million years. So I added the longest time-scale, infinity or eternity, all the way back to the beginning and all the way forward to the end. Beyond a million years in either direction, we are no longer human. But a million years is a short time in the history of our planet. Beyond a million years, the slow rhythms of biological and geological evolution become visible.

Ten years is, as Shakespeare knew, the normal horizon of human activities, the time we take to educate a child, to launch a career, to establish a business. This is the time-scale on which we are competing against one another as individuals, the time-scale for getting married and raising a family. Ten years is also the outer limit of political predictability. In ten years governments change, political leaders rise and fall, empires collapse, wars and revolutions turn the world upside down.

Economics and technology are more predictable than politics. On a time-scale of ten years, the forces driving economic and technological development will not change radically. The main cause of economic tensions today is the unequal distribution of wealth and skill between rich and poor countries and between rich and poor segments of society. It is likely that economic dislocations will intensify over the next ten years, as the rich countries grow richer and the poor grow poorer. This process of intensifying inequality must somehow be reversed, but the means for reversing it are not yet in sight. To reverse it, we shall need new political institutions as well as new technologies. New institutions and new technologies take longer than ten years to grow.

While economics and technology are slow to change, pure science is quick. In science we are accustomed to see radical changes happen in ten years or less. Ten years is a typical time-scale for a scientific revolution. The digital astronomy revolution that I described in Chapter Two is already half over. Within a few decades we will have digital sky surveys giving us maps of the whole observable universe. The younger astronomers alive today belong to a historically unique generation. Their generation is the first that will study cosmology with a detailed knowledge of the structure of the universe that they are trying to explain.

Another enterprise with a ten-year horizon is the Human Genome Project. The Genome Project will produce a precise digital map of our genes and chromosomes, like the maps of the sky produced by digital sky surveys. The maps will be used in similar ways, either for identifying particularly interesting objects, genes, or galaxies as the case may be, which are good candidates for a deeper investigation, or for collecting statistical information about the behavior of genes or galaxies in general. Digital sky surveys became cheap and practical as a result of the invention of the Charge-Coupled Device, or CCD. Without the CCD, the surveys would be too slow and too expensive to be of interest to astronomers. The Human Genome Project is still slow and expensive. The equivalent of the CCD for the genome has not yet been invented. The biologists are still living in the pre-CCD era. The human genome consists of about 3 billion base-pairs, and the existing chemical methods of sequencing cost between ten cents and a dollar per base-pair. Using existing methods, the sequencing of a single human genome will require a major investment of money and time and will not be completed for many years.

The biological equivalent of the CCD is likely to appear soon. In astronomy, the revolutionary change from photographic plates to CCD detectors was a change from the technology of chemistry to the technology of physics. In biology, the move from chemistry to physics has not yet happened. But the tools of physics are rapidly improving, and will soon be directly applied to the job of sequencing DNA. In theory, scanning tunneling microscopes and atomic force microscopes already have resolution good enough to make visible the sequence of bases in a strand of DNA. They still have to overcome the formidable practical problems of localizing and preparing the DNA without destroying it. The practical obstacles to physical sequencing of DNA will probably be overcome within ten or twenty years.

As soon as physical sequencing is possible at all, it will be cheaper and more rapid than chemical sequencing. The jump in economy and speed from chemical to physical sequencing will be at least as large as the jump from the photographic plate to the CCD. A physical sequencer is likely to read out base-pairs at a rate of a hundred per second, compared with the hundreds per hour that chemical methods can achieve. If these expectations are fulfilled, a single machine sitting on a table-top will produce a complete human genome sequence every year. The Genome Project will then be able to create a library of sequences of men and women with a variety of medical and genetic histories, and a library of sequences of other species from bacteria to chimpanzees and whales. But first we have to invent the biological equivalent of the CCD. I recommend this invention as a task for any ambitious young person who dreams of leading a scientific revolution.

On a time-scale of a hundred years, each of us as an individual is dead. To survive on a time-scale of a hundred years means to survive as a family, as a nation, as a school of science or art, as an industrial enterprise, or as a religious community. Because our species has evolved to survive on a time-scale of hundreds of years, we have loyalties to these larger units that transcend our individual lives. The loyalties to family, tribe, and institution are deeply ingrained in our nature. Without these passionate loyalties to causes larger than ourselves, we would not be human. When we look to the future on a time-scale of a hundred years, the central question that we must ask is: to which causes shall our loyalties be directed?

A hundred years is the outer limit of technical predictability. Any single technology, such as coal, steam, electricity, computer chips, or recombinant DNA, remains dominant for at most a hundred years before being eclipsed by its successor. We can predict with some degree of assurance the dominant technologies of the next fifty years. My own prediction for this period would include the present dominant technologies of petroleum, computers, and biochemistry, plus the two newcomers, genetic engineering and artificial intelligence. A hundred years in the future, genetic engineering and artificial intelligence will be mature and ready to be superseded by something else, perhaps by radio-telepathy. We cannot predict what the new technologies beyond a hundred years will be, because they will be based on scientific discoveries not yet made.

It is likely that during the next hundred years our ecological battles will intensify and will come to dominate the political agenda of mankind. During this time we will be forced to confront the great problems of human overpopulation, destruction of natural ecologies, and economic inequality. Nobody today can predict how these problems will be solved, if they are solved.

I venture only two general predictions. First, the

successful solutions will be local rather than global, tailored to the needs and traditions of local populations. Solutions that are globally imposed are unlikely to be durable. Different parts of the globe have profoundly different views about the proper balance of individual rights and social obligations. People who try to impose global solutions should remember the words of the poet William Blake, "One Law for the Lion and Ox is Tyranny."

My second prediction about the problems of the next century is that the new technology of genetic engineering will change the nature of the problems and will make new solutions possible. Already in the present century we have seen a splendid example of a simple biological invention, the birth control pill, allowing populations in many countries to become stabilized without any coercive measures imposed by governments. In the coming century we will see more such examples, for good or for evil. Genetically engineered crop plants adapted to harsh environments and resistant to disease will transform the economy of nations. Genetically engineered babies, guaranteed to be free of hereditary defects, will be available to parents who are citizens of rich countries or to rich citizens of poor countries. New medical technologies, too attractive to forbid and too expensive to be made generally available, will exacerbate the inequalities that now exist within and between societies.

One more prediction can safely be made. A hundred years from now, our planet will not be a peaceful utopia with all its social and ecological problems solved. Even if economic inequalities can be greatly reduced, racial and religious animosities will persist. Another new technology, the technology of space colonization, may make it possible in the long run to alleviate the conflicts between discordant human ambitions on a shrinking planet. If it were possible to export surplus populations of people and industries into space habitats scattered over the solar system, then the earth might be preserved as an unspoiled wilderness or as an ecological park. The people who choose to remain on Earth might be required to live frugally and encroach no further on the living-space of other species. Polluting industries might be permanently banned and human populations on Earth might be firmly limited. For such utopian visions to become real, the technologies of space transport and space habitation must become radically cheaper and more accessible than they are today. Traveling and living in space must become cheap enough, so that ordinary people can emigrate from Earth as they now emigrate from one country to another. Political and social

PAUL BUNYAN'S TEAM OF OXEN WITH THE DRIVER

By Fred Smith, mid-twentieth century. Photograph by Eric Sutherland. Reproduced by permission of Friends of Fred Smith.

obstacles to emigration may still exist, but the physical and economic obstacles can be overcome if the technology is cheap enough.

I have no doubt that cheap space travel will sooner or later be developed. The question that I am not able to answer is when this will happen. What is the appropriate time-scale for the great migration from Earth? There is no law of physics or chemistry that decrees that space travel must always be expensive. So far as the laws of physics are concerned, if one measures the cost of moving into space by calculating the cost of the energy required, the cost of launching a person from Earth into space should be no greater than the cost of a commercial flight from New York to Tokyo. But to bring the cost of space launches down to the cost of commercial air travel requires a huge volume of traffic. Space travel will only be cheap when millions of people can use it. And millions of people can only use it when there are abundant space habitats already built, places to which the travelers can go. The growth of space habitats and the decline of costs must inevitably be a slow process. Much may be done in a hundred years, but not enough to have a major impact on the problems of Earth-bound humanity. The large-scale expansion of life and humanity into space will not come in time to solve the problems of our grandchildren.

In the next hundred years we will probably have human settlements on the Moon and on a few nearby asteroids. Perhaps also on Mars. We will see genetically engineered plants and animals adapted to the colonization of various asteroids and planets. For example, we

MARS COLONY

By unnamed Japanese artist, 1990s. Reproduced by permission of Obayashi Corporation.

may see the Martian potato established on Mars, a variety of potato that lives in places where liquid water lies deep under the frozen ground, with hardy shoots climbing up to the surface and sprouting leaves to take advantage of mid-day sunshine during the Martian summer. These developments will be necessary precursors of large-scale emigration but will not be sufficient. My guess is that large-scale emigration will begin only after many ventures of small-scale emigration have been attempted, some succeeding and others failing. Small-scale emigration may continue for a few hundred years before life is thoroughly adapted and growing wild on the multitude of worlds that are orbiting around the sun. Long before a thousand years have passed, life will have spread over the solar system. And by that time, our human descendants will be spread out too.

⚑

On a time-scale of a thousand years, neither politics nor technology is predictable. China and Japan are the only major political units that have lasted that long. A thousand years ago, Europe was an unimportant peninsula lying on the edge of the more advanced and civilized Arab world. The technologies of today would be unintelligible to our ancestors of a millennium ago. The only human institutions that retain their identities over a thousand years are languages, cultures, and religions. Perhaps it is not coincidental that the most intractable human quarrels and the most imperishable artistic creations are alike rooted in our languages, our cultures, and our religions.

Looking ahead a thousand years, one may predict that the diversity of languages, cultures, and religions will still exist, even if the dominant varieties are different from those that prevail today. The spreading out of human settlements into far distant places will tend to preserve our diversity and at the same time to make our diversity less dangerous. On a time-scale of a thousand years, the genetic differences between human populations may be increased by effects of natural selection or genetic engineering. Genetic differences which would be socially divisive and politically intolerable on Earth may be harmless when the deviant populations are living on distant asteroids. Within a thousand years, our descendants may be so widely dispersed that no central authority will be able to regulate their activities or even be aware of their existence. The process of speciation, the division of our species into many varieties with genetic endowment drifting gradually further apart, will then be under way. In the history of life on Earth, diversification of life forms has led to speciation, and speciation has led to further diversification. As humanity expands its living space away from the earth, the same processes are likely to occur. Our one species will become many. There is no reason why a variety of intelligent species should not fill a variety of ecological niches in different physical environments, some adapted to heat, others to cold, some to zero gravity, others to strong gravity, some to high pressure, others to living in the vacuum of space.

The main difference between the processes of natural speciation and the formation of human species in the future is a difference of time-scale. Speciation in nature occurs with a time-scale of the order of a million years. Human speciation pushed by genetic engineering may have a time-scale of a thousand years or less. Compared with the slow pace of natural evolution, our technological evolution is like an explosion. We are tearing apart the static world of our ancestors and replacing it with a new world that spins a thousand times faster.

Consider the simple question of size. The size of our population, the size of our economic resources, the size of our living space, all are growing at an average rate of about two percent per year. So far as the population on the earth is concerned, this two-percent growth must soon come to an end. But when life and industrial activities are spread out over the solar system, there is no compelling reason for growth to stop. It could happen that the growth will continue at a rate of two percent a year for a thousand years. Then, at the end of a thousand years, our population and resources and living space will have grown by a factor of five hundred million. There will still be ample reserves of sunlight and water and other essential materials available in the solar system to support a population of this size. I am not saying that it is necessary or desirable to increase our numbers by a factor of five hundred million. I am saying only that it is possible, and that it may happen within the next thousand years if we do nothing to prevent it.

More important than the growth in the quantity of human beings is the possibility of radical changes in quality. During the next thousand years there will be many opportunities for experiments in the radical reconstruction of human beings. Some of these experiments may succeed. When they succeed, our descendants may be born with mental qualities different from ours. To explore the possibilities of mental experience will be as great a challenge as the exploration of the physical universe. We must expect that at least some of our descendants will be eager to explore the delights of collective memory and collective consciousness, made possible by the technology of radiotelepathy. The experience of collective memory and collective consciousness will enormously enlarge the scope of art, science, religion, and history. Other experiments in collective consciousness may link human brains with those of dolphins and whales, lions and chimpanzees and eagles, breaking down barriers not only between individuals but between species. Those who have experienced the merging of memory and consciousness into a larger mind may find it difficult to communicate with those who still rely on spoken or written words. Those who have been part of an immortal group-mind may find it difficult to communicate with ordinary mortals.

The most serious conflicts of the next thousand years will probably be biological battles, fought between different conceptions of what a human being ought to be. Societies of collective minds will be battling against societies of old-fashioned individuals. Big brains will be battling against little brains. Devotees of artificial intelligence will be battling against devotees of natural wisdom. Such battles may lead to wars of genocide. But the vast expanses of space beyond the earth offer a way to resolve such biological battles peacefully. Within a thousand years, life will have spread through the solar system to the outer reaches of the Kuiper belt of comets, a thousand times the earth's distance from the sun. Societies that disagree fundamentally concerning the meaning and purpose of life may agree to keep out of each other's way by migrating to opposite ends of the solar system. Space is big enough to have room for them all.

We may hope that one group of our descendants, those who cling to our old human heritage, those who are loyal to our natural human shape and genetic endowment, will be allowed to remain here in possession of our planet, to maintain the old human values in our original birthplace, while those who radically transform themselves into new shapes will move away out of sight and out of reach.

If life succeeds in escaping from Earth and spreading out into the universe, the next thousand years might be a golden age of science. Voyages of exploration might already be moving out beyond the solar system to interstellar distances. And voyages of mental exploration might be moving out in many directions that we cannot yet imagine. We do not know even the names of the new sciences that may rise and fall before a thousand years have gone by.

On a time-scale of ten thousand years, the mismatch between our past and our future becomes even more acute. Ten thousand years ago we were still a single species, not noticeably different in physical and mental qualities from the people of today. We were living in hunter-gatherer societies, and learning slowly how to adapt to a warmer climate after the rapid ending of the last ice age. Our loyalties were fixed to family, tribe, and local culture. Ten thousand years in the future, who knows what we shall be? On the ten-thousand-year time-scale, qualitative changes dominate quantitative changes. On that time-scale, our values and ideals are totally plastic. The battle-ground of human evolution will move from biology to philosophy. Science may or may not still exist. Beings that we would recognize as human may or may not still exist. I hope that our human shape and our ancient human loyalties will be preserved in some fraction of our future territory. Even if our descendants in other regions have achieved immortality, as they well may, it would be wise to keep a population of mortal humans on Earth, so that some contact with the reality of death will not be lost. Our descendants may, as Olaf Stapledon imagined, refresh their spirits with a "Cult of Evanescence," a form of religion or artistic creation in which the tragedy and beauty of short-lived creatures is given the highest value. The cult of evanescence may be an anchor, connecting a computerized and intellectualized species with the ancient realities of life and death.

In the long run, the central problem of any intelligent species is the problem of sanity. We shall be free to choose our values and our purposes. There will be no absolute standards by which to judge one set of values right and another wrong. There will be ever-present the danger that a society will fall into a vicious circle, trapped in a dream world of its own devising by a value system that has lost contact with reality. A society trapped in this way is like a drug addict whose value system the drug has short-circuited. For the addict, the drug is a driving force more powerful than any of the normal human values. For a society with a technological control of human emotions, addictions to artificial emotional experiences may be fatally easy to induce. A society addicted in this way to dreams and shadows has lost its sanity. It is a danger to itself and to others.

The only cure for an insane society is harsh contact with reality. Sanity is the ability to live in harmony with Nature's laws. To remain sane, our descendants must strive to keep the emotional roots of our species intact, to preserve the emotional balance that we evolved through millions of years of living precariously in a natural environment. If we are to survive through a long future, we must stay in contact with our long past. It is not only for aesthetic reasons that we should preserve the earth as a cultural museum. The earth with its millions of species will offer to our descendants an object-lesson in the art of living. It will give them a reality check which they will need more and more, the further they move away from it.

⟁

A hundred thousand years ago, we were busy learning how to be human. We were adapting ourselves to a cold climate with clever inventions and abstract language and foresight. We were beginning to educate our children. We were learning a fierce loyalty to our own species in competition with our Neanderthal cousins. A hundred thousand years in the future, nothing is predictable. By then, we could have spread life all over our galaxy, if life is not spread over the galaxy already. We will have made contact with whatever alien forms of life exist in the galaxy. If we are lucky, our history may be enriched by a multitude of alien cultures and traditions. The aliens will probably have notions of good and evil very different from ours. We will have much to teach and much to learn.

Perhaps, on a time-scale of a hundred thousand years, the dominant factor in life's history will be the adaptation to living in a Carroll universe. Jean-Marc Lévy-Leblond invented this universe as a mathematical exercise, showing that it is one of a small number of model universes that are logically self-consistent. Lévy-Leblond, a French physicist now living in Nice, named his universe after Lewis Carroll, the English mathematician who wrote the classic books for children, *Alice in Wonderland* and *Through the Looking-Glass*. In the looking-glass world of Carroll, the Red Queen says to Alice, "It takes all the running *you* can do, just to stay in the same place." So in Lévy-Leblond's Carroll universe, nothing ever moves from one place to another. Even light has zero velocity.

When life is spread all over the galaxy, we will be living in a Carroll universe because distances are too large to be traveled on a human time-scale. Even messages traveling at the speed of light take fifty thousand years to creep across the galaxy. Whole historical epochs will pass, cultures will rise and fall, between a telephone call and the reply. Each little piece of the galaxy will be a world of its own, isolated from other pieces by the immensity of space and the quickness of time. We shall enjoy abundant communication with our neighbors in the past, but of our neighbors in the present we can know nothing.

There are two other simple model universes that are more familiar than the Carroll universe. These are the Newton universe and the Einstein universe. In the Einstein universe, both space and time are relative. In the Newton universe, time is absolute and space is relative. In the Carroll universe, space is absolute and time is relative. It is a curious paradox that we lived for a hundred thousand years in the past in a Carroll universe, and shall live in the future in a Carroll universe, with only a short interval of Newton and Einstein in between.

In the past, before the invention of ships and wheels, we lived in a Carroll universe because each little tribe of humans could move only a short distance within a human lifetime. Each tribe was like a point in a Carroll universe, separated from other points by absolute space. Then, when ships and wheels were invented, we learned to travel around the world. Our space was no longer absolute and we moved into a Newton universe. A little later we invented the telegraph and the radio and moved into an Einstein universe. We will stay in an Einstein universe for a few thousand years, until we spread out over interstellar distances or contact aliens who are already spread out over the galaxy. And then, after we are spread out, we shall be back in the Carroll universe. In the long future, it may be helpful to us that we still carry the traces of our long past in a Carroll universe. The Carroll universe made us what we are, a territorial species with intense loyalties to place and home. In the Newton and Einstein universes of the recent past and present, these loyalties became dangerous and destructive, driving us to territorial wars and extermination of peoples. In the bigger Carroll universe of our future, we shall no longer be capable of physically attacking our neighbors, and the ancient territorial loyalties will become once again benign.

A few years ago, I was asked by OMNI magazine to compose a message, in seventy-five words or less, to be sent by radio to any aliens who might be listening. Here is the message that I would send:

> Dear Aliens, your silence puts us to shame. Please forgive us for making so much noise in this beautiful universe which we are sharing with you. Please be patient when we are impatient, be gentle when we are rough, be wise when we are stupid. We are a young species and still have much to learn.

⚜

A million years in the past takes us back to the beginning of our species. During these million years we made all the great inventions that stamp us as human. Language, to share our joys and our sorrows, our thoughts and our skills. Grandparents, the third generation that we added to the mammalian family, to give us leisure to educate children and spin stories around the campfire while the parents are busy hunting and gathering. Awareness of death, to enlarge our horizons in time and reveal to us our place as links in a chain of being, revering our ancestors and cherishing our descendants. Laughter, to enable us to enjoy the absurdities of our situation. Love of sunshine and rivers and trees and earth. Reverence and loyalty, not only to our own tribe but to other species and to Nature as a whole.

Religion, adding a mystical dimension to give meaning to our struggles. And finally, among other deities, the Earth Goddess Gaia who symbolizes the nurturing and protecting power of our planet.

Going a million years into the future, we shall make new inventions at least as profound and revolutionary as those that we made in the past. We today can have no inkling of the nature of our future inventions. The concerns of our descendants a million years in the future would probably be as unintelligible to us today as differential equations or astrophysics would have been to an early hominid roaming the plains of Africa. All that we can say about the future a million years ahead is quantitative rather than qualitative. We cannot describe the quality of our descendants' lives, but we can predict roughly where they will be. They will be spread over our galaxy from end to end and will be reaching out toward other galaxies. They will be directly aware, with a depth of understanding that we cannot imagine, of the entire history of that part of the universe that is within their past horizon. They will be sending out communications to their distant neighbors in all that part of the universe that is within their future horizon. In a Carroll universe, the gap between the past and future horizons is wide. Even for a neighbor in a nearby galaxy such as M31, the gap between the two horizons, the gap between the communicable past and the communicable future, is several million years. That is what it means to live in a Carroll universe. A million years from now, our descendants and their neighbors in other galaxies will perhaps be preparing for the intelligent intervention of life in the evolution of the universe as a whole. That is an adventure whose beginning we can conjecture, but from here it is out of sight.

Beyond a million years into the past, we are no longer human. But life has a history going back three billion years, and that history is also ours. I do not attempt here a detailed sketch of life's history. I mention only one aspect of that history that is relevant to life's long-range future, namely the concept of Gaia. Gaia is not only the name of an old Greek goddess, it is also the name of a scientific theory having nothing to do with religion or with mysticism. The scientific Gaia is a theory of the history of life on Earth, first propounded by James Lovelock. The theory asserts that the chemistry and ecology of Earth are linked in a single system that keeps the environment of the planet within limits tolerable to life. Gaia means simply the control system that ties together the actions of life and the environment. I am not expert in planetary science, but I find the Gaia theory plausible. It describes the way Earth and its inhabitants interact. It does not claim to explain how the interactions work.

As an example to illustrate the scientific evidence for Gaia, I choose the concentration of salt in seawater. Lovelock has collected geological evidence showing that the salt in the oceans has remained at a roughly constant concentration for the last two billion years. On the other hand, the quantities of salt deposited into the oceans by rivers would be enough to double the salt concentration in about a hundred million years. If some process of regulation had not been holding the concentration constant, the oceans would by now be as saturated with salt as the Dead Sea, and almost all forms of marine life would be extinct. By some mechanisms not yet understood, the system that Lovelock calls Gaia has limited the salt concentration and has kept the oceans fresh enough to be a comfortable home for fish and other creatures. One mechanism for removing salt from the oceans is the deposition of solid salt in shallow lagoons. But nobody yet understands how this process is regulated so as to hold the dissolved salt constant over two billion years.

Although the processes of regulation are not understood, we know that Gaia somehow maintained this planet as a home for life, in spite of many climatic changes and geological jolts, for three billion years. We know that if we wish to continue living on Earth in health and comfort, we must keep Gaia healthy too. Our future on Earth depends on understanding and preserving Gaia. Even before we were human, our bodies and minds were evolved to live in equilibrium with Gaia. In the future beyond a million years, after we cease to be human, we shall still survive in equilibrium with Gaia if we survive at all.

Beyond a million years, life will go on, but we cannot hope to understand what life will be doing, any more than a dinosaur could understand what we are doing now. We may hope that as life expands through the galaxy and through the universe, Gaia will expand too. The greening of the galaxy, or the greening of the universe, will be a slow process, following the slow rhythms of Gaia rather than the quick rhythms of impatient humans. Life as it expands must carry with it the mechanisms of self-regulation that keep it from uncontrolled growth and self-destruction. Life as it spreads must carry with it the constraints of Gaia. We may imagine that, in the end, a universal Gaia will be regulating life in every corner of the cosmos.

One of my favorite science-fiction stories is "Dragon's Egg," written by Robert Forward, who is a successful engineer as well as a gifted writer. "Dragon's Egg" is a story about an encounter between two societies living on different time-scales, our human society on Earth encountering a society of alien creatures who call themselves Cheelas, living on the surface of a neutron star. The Cheelas live a thousand times faster than we do. A day of their time is a minute of our time. A day of our time is several years of their time. The story hinges on the question: how can two societies with such different time-scales communicate with each other? In the story, the communication happens only once, very briefly, and then the two societies go their separate ways. This is not a satisfactory answer to the question. In fact, as the story makes clear, we do not know the answer.

We are faced with the same question when we consider the future of human society on Earth. Robert Forward's story may be read as an allegory of short-lived humans getting acquainted with long-lived Gaia. We have to learn how to fit our lives and our destiny into the life and destiny of Gaia. The fundamental difficulty here is the difference in time-scales, the time-scales of human life being ten or a hundred years, the time-scales of Gaia being a million or a billion years. Somehow we have to co-exist and cooperate with Gaia across this immense barrier of time-scales.

Our problem is similar to the problem of the Cheelas. From the Cheela point of view, humans are diffuse and poorly defined. We are also dumb and slow. We are so slow that we drive them crazy with impatience. From the human point of view, Gaia is diffuse, poorly defined, dumb, and slow. Her slowness drives us crazy. That is one of the central facts of the human predicament. We have to learn the art of living on Gaia's time-scale as well as our own.

❦

We cannot predict the evolution of life beyond a million years. We can only study the limits set to life's potentialities by the laws of physics and information theory. Fifteen years ago I published such a study, assuming as a basis for my calculations a model universe of subcritical density. Subcritical density means that the universe contains less matter than is required to halt the process of universal expansion. In a subcritical universe, the galaxies continue moving away from one another forever with constant velocities. I was able to show that in a subcritical universe, although the resources of matter and energy available in each galaxy would be finite, the laws of physics and information theory allow life to survive forever using a finite store of energy.

There are two possible alternatives to the subcritical universe. The density may be supercritical or precisely critical. In a supercritical universe, the density of matter is large enough to reverse the process of universal expansion within a finite time, and within another finite time the universe contracts to a state of infinite density. The universe which began its existence in a "big bang" ends its existence in a "big crunch." In a supercritical universe, the history of life is ended by quickly rising temperatures shortly before the big crunch. This is a dismal prospect, and I will say no more about it.

Fortunately, the possibility still exists that our universe may be precisely critical. In a precisely critical universe, the universal expansion is never reversed, but the distant galaxies move away from us more and more slowly as time goes on. Stephen Frautschi has published a study of the long-range future of life in an exactly critical universe. The critical universe is much friendlier to life than the subcritical universe. In the subcritical universe, each galaxy is finally alone, dependent on its own resources, with only radio communication connecting it with its rapidly receding neighbors. In the critical universe you can reach out and touch your neighbors. The neighboring galaxies are moving so slowly that you can visit and exchange material resources as well as information. You can cooperate with your neighbors in large-scale engineering projects to keep the universe in trim and maintain the optimum conditions for life. In a critical universe, the universal Gaia has far greater scope for her activities.

We have followed the evolution of life on seven different time-scales, going all the way from ten years to infinity. The one theme that is heard all through the story is that life brings drama and excitement to an otherwise dead and dull universe. At the end we see that the drama of life reaches a universal scale only in the critical universe, the universe that is for ever teetering on the edge between expansion and collapse. If it should turn out that the universe we are living in is actually critical, this would be strong evidence in support of a philosophical principle which I call the principle of maximum diversity.

The principle of maximum diversity states that the laws of Nature are constructed in such a way as to make the universe as interesting as possible. Most of the new discoveries in science during my lifetime have strengthened my belief that this principle is true. But the evidence for a critical universe is not as strong as we might wish it to be. The evidence at present tends to favor a subcritical universe. As a scientist, I am allowed to wish for a critical universe, but I am not allowed to confuse wishes with facts.

chapter five

ETHICS

CRUCIFIXION OF LABOR

By S. P. Dinsmoor, early twentieth century. Photograph by Eric Sutherland. Reproduced by permission of the Walker Art Center.

ONE OF MY FAVORITE MONUMENTS IS
a statue of Samuel Gompers not far from the Alamo in
San Antonio, Texas. Under the statue is a quotation
from one of Gompers's speeches:

> What does labor want?
> We want more school houses and less jails,
> More books and less guns,
> More learning and less vice,
> More leisure and less greed,
> More justice and less revenge,
> We want more opportunities to cultivate our
>   better nature.

To the Doers
Dedicated September 6, 1982
Sculptor Bette Jean Alden

I do not know how many of the thousands of tourists who come every year to San Antonio to pay their respects to Davy Crockett and the other heroes of the Alamo take a few moments on the way to listen to Samuel Gompers. I hope some of them find, as I did, the peaceful words of Gompers a fitting antidote to the cult of military madness symbolized by the Alamo. It comes as a refreshing surprise to hear, so close to the heart of Texas, so close to the shrine of patriotic pride, a quiet voice of reason.

Samuel Gompers was the founder and first president of the American Federation of Labor. He was largely responsible for the fact that the American labor movement broke away from the European movement dominated by the ideology of Karl Marx. The European labor leaders dreamed of a proletarian revolution. Gompers understood that American working people were not interested in revolution and dreamed mostly of high wages and economic security. Gompers was the champion of pragmatism against ideology. As a general rule, technologies driven by pragmatism work well, and technologies driven by ideology work badly. The

STATUE OF SAMUEL GOMPERS AT THE ALAMO

Photograph by Barbara Happer, 1996. Reproduced by permission of the photographer.

life of Gompers illustrates another general rule, that social justice fits better with pragmatism than with ideology.

It is an ironic fact of history that now, seventy years after his death, the ideas of Gompers have triumphed in Europe and failed in the United States. In Gompers's lifetime the revolutionary ideologues of Europe led their unions into disaster after disaster, the worst disaster being the dictatorship of the proletariat in Russia in 1917. Meanwhile, Gompers established in America the tradition of practical bargaining between labor and management which led to an era of growth and prosperity for the unions. Then, after the devastations of World War II, the positions on the two sides of the Atlantic Ocean were gradually reversed. The Europeans recognized the wisdom of Gompers and rebuilt their societies on a nonideological basis, the unions sharing power and responsibility for economic decisions.

Meanwhile, the United States forgot Gompers and embraced an ideology of doctrinaire free-market capitalism. In America the unions dwindled, while Gompers's dreams—more books and fewer guns, more leisure and less greed, more schoolhouses and fewer jails—were tacitly abandoned. In a society without social justice and with a free-market ideology, guns, greed, and jails are bound to win.

This book is supposed to be about technology, not about social justice, but you cannot tell which technology is good and which is bad without paying some attention to social justice. When I was a student of mathematics in England fifty years ago, one of my teachers was the great mathematician G. H. Hardy, who wrote a little book, *A Mathematician's Apology*, explaining to the general public what mathematicians do. Hardy proudly proclaimed that his life had been devoted to the creation of totally useless works of abstract art, without any possible practical application. He had strong views about technology, which he summarized in the statement, "A science is said to be useful if its development tends to accentuate the existing inequalities in the distribution of wealth, or more directly promotes the destruction of human life." He wrote these words while war was raging around him. Still, the Hardy view of technology has some merit even in peacetime. Many of the technologies that are now racing ahead most rapidly, replacing human workers in factories and offices with machines, making stock-holders richer and workers poorer, are indeed tending to accentuate the existing inequalities in the distribution of wealth. And the technologies of lethal force continue to be as profitable today as they were in Hardy's time. The marketplace judges technologies by their practical effectiveness, by whether they succeed or fail to do the job they are designed to do. But always, even for the most brilliantly successful technology, an ethical question lurks in the background, the question whether the job the technology is designed to do is actually worth doing.

The technologies that raise the fewest ethical problems are those that work on a human scale, brightening the lives of individual people. Lucky individuals in each generation find technology appropriate to their needs. For my father ninety years ago, technology was a motorcycle. He was an impoverished young musician growing up in England in the years before World War I, and the motorcycle came to him as a liberation. He was a working-class boy in a country still dominated by the snobberies of class and accent. He learned to speak like a gentleman, but he did not belong in the world of gentlemen. The motorcycle was a great equalizer. On his motorcycle, he was the equal of a gentleman. He could make the grand tour of Europe without having inherited an upper-class income. He and three of his friends bought motorcycles and rode them all over Europe.

My father fell in love with his motorcycle and with the technical skills that it demanded. He understood, sixty years before Robert Pirsig wrote *Zen and the Art of Motorcycle Maintenance* the spiritual quality of the motorcycle. In my father's day, roads were bad and repair shops few and far between. If you intended to travel any long distance, you needed to carry your own tool kit and spare parts and be prepared to take the machine apart and put it back together again. A breakdown of the machine in a remote place often required major surgery. It was as essential for a rider to understand the anatomy and physiology of the motorcycle as it was for a surgeon to understand the anatomy and physiology of a patient. It sometimes happened that my father and his friends would arrive at a village where no motorcycle had ever been seen before. When this happened, they would give rides to the village children and hope to be rewarded with a free supper at the village inn. Technology in the shape of a motorcycle was comradeship and freedom.

Fifty years after my father, I discovered joyful technology in the shape of a nuclear fission reactor. That was in 1956, in the first intoxicating days of peaceful nuclear energy, when the technology of reactors suddenly emerged from wartime secrecy and the public was invited to come and play with it. This was an invitation that I could not refuse. It looked then as if nuclear energy would be the great equalizer, providing cheap and abundant energy to rich and poor alike, just as fifty years earlier the motorcycle gave mobility to rich and poor alike in class-ridden England.

I joined the General Atomic Company in San Diego where my friends were playing with the new technology. We invented and built a little reactor which we called the TRIGA, designed to be inherently safe. Inherent safety meant that it would not misbehave even if the people operating it were grossly incompetent. The company has been manufacturing and selling TRIGA reactors for forty years and is still selling them today, mostly to hospitals and medical centers, where they produce short-lived isotopes for diagnostic purposes. They have never misbehaved or caused any danger to the people who used them. They have only run into trouble in a few places where the neighbors objected to their presence on ideological grounds, no matter how safe they might be. We were successful with the TRIGA because it was designed to do a useful job at a price that a big hospital could afford. The price in 1956 was a quarter of a million dollars. Our work with the TRIGA was joyful because we finished it quickly, before the technology became entangled with politics and

**GEORGE B. DYSON WITH HIS CAD-CAM
MACHINES**

*Photograph by Ann E. Yow, 1996. Reproduced by permission
of the photographer.*

bureaucracy, before it became clear that nuclear energy
was not and never could be the great equalizer.

Forty years after the invention of TRIGA, my son
George found another joyful and useful technology, the
technology of CAD-CAM, computer-aided design and
computer-aided manufacturing. CAD-CAM is the
technology of the postnuclear generation, the technology that succeeded after nuclear energy failed. George

is a boat-builder. He designs sea-going kayaks. He uses modern materials to reconstruct the ancient craft of the Aleuts, who perfected their boats by trial and error over thousands of years and used them to travel prodigious distances across the northern Pacific. His boats are fast and rugged and sea-worthy. When he began his boat-building twenty-five years ago, he was a nomad, traveling up and down the north Pacific coast, trying to live like an Aleut and build his boats like an Aleut, shaping every part of each boat and stitching them together with his own hands. In those days he was a nature-child, in love with the wilderness, rejecting the urban society in which he had grown up. He built boats for his own use and for his friends, not as a commercial business.

As the years went by George made a graceful transition from the role of rebellious teenager to the role of solid citizen. He married, raised a daughter, bought a house in the city of Bellingham, and converted an abandoned tavern by the waterfront into a well-equipped workshop for his boats. His boats are now a business. And he discovered the joys of CAD-CAM.

His workshop now contains more computers and software than sewing needles and hand tools. It is a long time since he made the parts of a boat by hand. He now translates his designs directly into CAD-CAM software and transmits them electronically to a manufacturer who produces the parts. George collects the parts and sells them by mail-order to his regular customers with instructions for assembling them into boats. Only on rare occasions, when a wealthy customer pays for a custom-built job, George delivers a boat assembled in the workshop. The boat business occupies only a part of his time. He also runs a historical society concerned with the history and ethnography of the north Pacific. The technology of CAD-CAM has given George resources and leisure, so that he can visit the Aleuts in their native islands and reintroduce to the young islanders the forgotten skills of their ancestors.

Forty years into the future, which joyful new technology will be enriching the lives of our grandchildren? Perhaps they will be designing their own dogs and cats. Just as the technology of CAD-CAM began in the production lines of large manufacturing companies and later became accessible to individual citizens like George, the technology of genetic engineering may soon spread out from the biotechnology companies and agricultural industries and become accessible to our grandchildren. Designing dogs and cats in the privacy of a home may become as easy as designing boats in a waterfront workshop.

Instead of CAD-CAM we may have CAS-CAR,

computer-aided selection and computer-aided reproduction. With the CAS-CAR software, you first program your pet's color scheme and behavior, and then transmit the program electronically to the artificial fertilization laboratory for implementation. Twelve weeks later, your pet is born, satisfaction guaranteed by the software company. When I recently described these possibilities in a public lecture at a children's museum in Vermont, I was verbally assaulted by a young woman in the audience. She accused me of violating the rights of animals. She said I was a typical scientist, one of those cruel people who spend their lives torturing animals for fun. I tried in vain to placate her by saying that I was only speaking of possibilities, that I was not actually myself engaged in designing dogs and cats. I had to admit that she had a legitimate complaint. Designing dogs and cats is an ethically dubious business. It is not as innocent as designing boats.

When the time comes, when the CAS-CAR software is available, when anybody with access to the software can order a dog with pink and purple spots that can crow like a rooster, some tough decisions will have to be made. Shall we allow private citizens to create dogs who will be objects of contempt and ridicule, unable to take their rightful place in dog society? And if not, where shall we draw the line between legitimate animal breeding and illegitimate creation of monsters? These are difficult questions that our grandchildren will have to answer. Perhaps I should have spoken to the audience in Vermont about designing roses and orchids instead of dogs and cats. Nobody seems to care so deeply for the dignity of roses and orchids. Vegetables, it seems, do not have rights. Dogs and cats are too close to being human. They have feelings like ours. If our grandchildren are allowed to design their own dogs and cats, the next step will be using the CAS-CAR software to design their own babies. Before that next step is reached, they ought to think carefully about the consequences.

Software for designing babies is still a small cloud on a distant horizon. It may well dissipate and disappear before it can come to disturb our peace. Meanwhile we have other more urgent questions to answer. What should be done to stop the destruction of forests and the extinction of species? What should be done to curb unsustainable growth of human populations? What should be done to dispose of the tens of thousands of nuclear weapons still defiling our planet after the Cold War is over? I do not have answers to these questions. The answers will have more to do with ethics and politics than with science and technology. I have expert knowledge of only one of these questions, the problem of nuclear weapons. A few years ago I walked into a room where there were forty-two hydrogen bombs lying around on the floor, not even chained down, each of them ten times as powerful as the bomb that destroyed Hiroshima. This experience was a sharp reminder of the precariousness of the human condition. It encouraged me to think hard about ways to improve the chances of survival of my grandchildren. Nuclear weapons remain, as George Kennan has said, the most serious danger to mankind and the most serious insult to God.

In the first four chapters of this book, nuclear weapons are hardly mentioned. The disappearance of nuclear weapons from our thinking about the future is a historic change for which we must be profoundly grateful. Fifty years ago and for many years thereafter, nuclear weapons dominated the landscape of our fears. The nuclear arms race was the central ethical problem of our age. Discussion of the ethical dilemmas of scientists centered around bombs and long-range missiles. The evil face of science was personified by the nuclear bomb designer. Now, quietly and unexpectedly, the bombs have faded from our view. But they have not ceased to exist. The danger to humanity of huge stockpiles in the hands of unreliable people is as real as ever.

Yet the bombs are not mentioned in our vision of the future. How could this have happened?

In the summer of 1995 I took part in a technical study of the future of the United States nuclear stockpile. The study was done by a group of academic scientists together with a group of professional bomb designers from the weapons laboratories. The purpose of the study was to answer a question. Would it be technically feasible to maintain forever a stockpile of reliable nuclear weapons of existing designs without further nuclear tests? The study did not address the underlying political questions, whether reliable nuclear weapons would always be needed and whether further nuclear tests would always be undesirable. Each of us had private opinions about the political questions, but politics was not the business of our study. We assumed as the ground rule for the study that the weapons in the permanent stockpile must be repaired and remanufactured without change in design as the components deteriorate and decay. We assumed that the new components would differ from the old ones when replacements were made, because the factories making the old components would no longer exist. We looked in detail at each type of weapon and checked that its functioning was sufficiently robust so that minor changes in the components would not cause it to fail. We concluded our study with a unanimous report, saying that a permanently reliable nuclear stockpile without nuclear testing is feasible. Unanimity was essential.

Unanimity was made possible by the objectivity and the personal integrity of the four weapons designers who worked side by side with us for seven weeks, John Richter and John Kammerdiener from Los Alamos, Seymour Sack from Livermore, and Robert Peurifoy from Sandia. They are impressive people, master craftsmen of a demanding technology. They have spent the best part of their lives planning and carrying out bomb tests. They remember every test, whether it succeeded or failed. They know why each test was done, and what was learned from its success or failure. Their presence was essential to our work, and their names on the report gave credibility to our conclusions. They are survivors of a vanishing culture. They lived through the heroic age of weapon-building. They will not and cannot be replaced. By working on this study, they unselfishly helped our country to move safely into a world in which people with the special qualities and talents of these four men will no longer be needed.

The conclusion of our study was a historical landmark, commemorating the fact that the nuclear arms race is finally over. The nuclear arms race raged with full fury for only twenty years, the 1940s and 1950s.

Then it petered out slowly for the next thirty years, in three stages. The science race petered out in the 1960s, after the development of highly efficient hydrogen bombs. Nuclear weapons then ceased to be a scientific challenge. The military race petered out in the 1970s, after the development of reliable and invulnerable missiles and submarines. Nuclear weapons then ceased to give a military advantage to their owners in real-world conflicts. The political race petered out in the 1980s, after it became clear to all concerned that massive nuclear weapons industries were environmentally and economically disastrous. The size of the nuclear stockpile then ceased to be a political status symbol. Arms control treaties were concluded at each stage, to ratify with legal solemnity the gradual petering out of the race. The atmospheric test ban ratified the end of the science race, the ABM and SALT treaties ratified the end of the military race, and the START treaties ratified the end of the political race.

How may we extrapolate this history into the world of the 1990s and beyond? The security and the military strength of the United States now depend primarily on nonnuclear forces. Nuclear weapons are on balance a liability rather than an asset. The security of the United States will be enhanced if all deployments of nuclear weapons, including our own, are gradually reduced to zero. For the next fifty years we should attempt to drive the nuclear arms race in reverse gear, to persuade our allies and our enemies that nuclear weapons are more trouble than they are worth. The most effective moves in this direction are unilateral withdrawals of weapons. The move that signaled the historic shift of the arms race into reverse gear was the unilateral withdrawal of land-based and sea-based tactical nuclear weapons by President Bush in 1991. Chairman Gorbachev responded quickly with similarly massive withdrawals of Soviet weapons. The testing moratorium of 1992 was another effective move in the same direction.

To drive the nuclear arms race further in reverse gear, we need to pursue three long-range objectives: worldwide withdrawal and destruction of weapons, complete cessation of nuclear testing, and an open world in which nuclear activities of all countries are to some extent transparent. In pursuing these objectives, unilateral moves are usually more persuasive than treaties. Unilateral moves tend to create trust, whereas negotiation of treaties often tends to create suspicion.

Our nuclear stockpile study fitted well into the context of the reverse-gear arms race. The purpose of the study was to achieve a technical stabilization of our stockpile, to clarify what needs to be done to maintain a limited variety of weapons indefinitely without testing. Stabilization is the essential prerequisite for allowing the weapons to disappear gracefully. Once a stable regime of stockpile maintenance has been established, the weapons will attract less attention both nationally and internationally. They will acquire the qualities that a stable nuclear deterrent force should have: awesomeness, remoteness, silence. Gradually, as the decades of the twenty-first century roll by, these weapons will become less and less relevant to the problems of international order in a hungry and turbulent world. The time may come when nuclear weapons are perceived as useless relics of a vanished era, like the horses of an aristocratic cavalry regiment, maintained only for ceremonial purposes. When nuclear weapons are generally regarded as absurd and irrelevant, the time may have come when it will be possible to get rid of them altogether.

The time when we can say good-by to nuclear weapons is still far distant, too far to be clearly envisaged, perhaps a hundred years away. Until that time comes, we must live with our weapons as responsibly and as quietly as we can. That was the purpose of our stockpile study, to make sure that our weapons can be maintained with a maximum of professional competence and a minimum of fuss and excitement, until in the fullness of time they will no longer be considered necessary. In the meantime, the ethical dilemmas concerned with nonnuclear weapons and nonnuclear warfare remain unresolved.

The abolition of war is an ultimate goal, more remote than the abolition of nuclear weapons. The idea espoused early in the nuclear age by Robert Oppenheimer, that the existence of nuclear weapons might lead to the abolition of war, turned out to be an illusion. The abolition of war is a prime example of an ethical problem that science is powerless to deal with. The weapons of nonnuclear war, guns and tanks and ships and airplanes, are available on the open market to anybody with money to pay for them. Science cannot cause these weapons to disappear. The most useful contribution that science can make to the abolition of war has nothing to do with technology. The international community of scientists may help to abolish war by setting an example to the world of practical cooperation extending across barriers of nationality, language, and culture.

⚉

Seventy-four years have passed since Haldane's *Daedalus* was published, and his warnings of science turning good into evil have been justified by events. The main question raised by Haldane's book is how the destiny that he assigns to Daedalus might be reversed. What can we do today, in the world as we find it at the end of the twentieth century, to turn the evil consequences of technology into good? The ways in which science may work for good and evil in human society are many and various. As a general rule, to which there are many exceptions, science works for evil when its effect is to provide toys for the rich, and works for good when its effect is to provide necessities for the poor. Cheapness is an essential virtue. The motorcycle worked for good because it was cheap enough for a poor schoolteacher to own. Nuclear energy worked mostly for evil because it remained a toy for rich governments and rich companies to play with. "Toys for the rich" mean not only toys in the literal sense but technological conveniences that are available to a minority of people and make it harder for those excluded to take part in the economic and cultural life of the community. "Necessities for the poor" include not only food and shelter but adequate public health services, adequate public transportation, and access to decent education and jobs.

The scientific advances of the nineteenth century and the first half of the twentieth were generally beneficial to society as a whole, spreading wealth to rich and poor alike with some degree of equity. The electric light, the telephone, the refrigerator, radio, television, synthetic fabrics, antibiotics, vitamins, and vaccines were social equalizers, making life safer and more comfortable for almost everybody, tending to narrow the gap between rich and poor rather than to widen it. Only in the second half of our century has the balance of advantage shifted. During the last forty years, the strongest efforts in pure science have been concentrated in highly esoteric fields remote from contact with everyday problems. Particle physics, low-temperature physics, and extragalactic astronomy are examples of pure sciences moving further and further away from their origins. The intensive pursuit of these sciences does not do much harm, or much good, either to the rich or the poor. The main social benefit provided by pure science in esoteric fields is to serve as a welfare program for scientists and engineers.

At the same time, the strongest efforts in applied science have been concentrated upon products that can be profitably sold. Since the rich can be expected to pay more than the poor for new products, market-driven applied science will usually result in the invention of toys for the rich. The laptop computer and the cellular telephone are the latest of the new toys. Now that a large fraction of high-paying jobs are advertised on the Internet, people excluded from the Internet are also excluded from access to jobs. The failure of science to produce benefits for the poor in recent decades is due to two factors working in combination: the pure scientists have become more detached from the mundane needs of humanity, and the applied scientists have become more attached to immediate profitability.

Although pure and applied science may appear to be moving in opposite directions, there is a single underlying cause that has affected them both. The cause is the power of committees in the administration and funding of science. In the case of pure science, the committees are composed of scientific experts performing the rituals of peer review. If a committee of scientific experts selects research projects by majority vote, projects in fashionable fields are supported while those in unfashionable fields are not. In recent decades, the fashionable fields have been moving further and further into specialized areas remote from contact with things that we can see and touch. In the case of applied science, the committees are composed of business executives and managers. Such people usually give support to products that affluent customers like themselves can buy. Only a cantankerous individual like Henry Ford, with dictatorial power over his business, would dare to create a mass market for automobiles by arbitrarily setting his prices low enough and his wages high enough so that his workers could afford to buy his product. Both in pure science and in applied science, rule by committee discourages unfashionable and bold ventures. To bring about a real shift of priorities, scientists and entrepreneurs must assert their freedom to promote new technologies that are more friendly than the old to poor people and poor countries. The ethical standards of scientists must change as the scope of the good and evil caused by science has changed. In the long run, as Haldane and Einstein said, ethical progress is the only cure for the damage done by scientific progress.

The nuclear arms race is over, but the ethical problems raised by nonmilitary technology remain. The ethical problems arise from three "new ages" flooding over human society like tsunamis. First is the Information Age, already arrived and here to stay, driven by computers and digital memory. Second is the Biotechnology Age, due to arrive in full force early in the next century, driven by DNA sequencing and genetic engineering. Third is the Neurotechnology Age, likely to arrive later in the next century, driven by neural sensors and exposing the inner workings of human emotion and personality to manipulation. These three new technologies are profoundly disruptive. They offer liberation from ancient drudgery in factory, farm, and office. They offer healing of ancient diseases of body and mind. They offer wealth and power to the people who possess the skills to understand and control them. They destroy industries based on older technologies and make people trained in older skills useless. They are likely to by-pass the poor and reward the rich. They will tend, as Hardy said eighty years ago, to accentuate the inequalities in the existing distribution of wealth, even if they do not, like nuclear technology, more directly promote the destruction of human life.

The poorer half of humanity needs cheap housing, cheap health care, and cheap education, accessible to everybody, with high quality and high aesthetic standards. The fundamental problem for human society in the next century is the mismatch between the three new waves of technology and the three basic needs of poor people. The gap between technology and needs is wide and growing wider. If technology continues along its present course, ignoring the needs of the poor and showering benefits upon the rich, the poor will sooner or later rebel against the tyranny of technology and turn to irrational and violent remedies. In the future, as in the past, the revolt of the poor is likely to impoverish rich and poor together.

The widening gap between technology and human needs can only be filled by ethics. We have seen in the last thirty years many examples of the power of ethics. The worldwide environmental movement, basing its

HANFORD TANK FARM

These million-gallon double-walled carbon steel tanks were built to hold high-level nuclear waste from the plutonium program at Hanford Reservation, Richland, Washington. Photograph by Robert del Tredici, 1984. Reproduced by permission of the photographer.

power on ethical persuasion, has scored many victories over industrial wealth and technological arrogance. The most spectacular victory of the environmentalists was the downfall of nuclear industry in the United States and many other countries, first in the domain of nuclear power and more recently in the domain of weapons. It was the environmental movement that closed down factories for making nuclear weapons in the United States, from plutonium-producing Hanford to warhead-producing Rocky Flats. Ethics can be a force more powerful than politics and economics. Unfortunately, the environmental movement has so far concentrated its attention upon the evils that technology has done rather than upon the good that technology has failed to do. It is my hope that the attention of the Greens will shift in the next century from the negative to the positive. Ethical victories putting an end to technological follies are not enough. We need ethical victories of a different kind, engaging the power of technology positively in the pursuit of social justice.

If we can agree with Thomas Jefferson that these truths are self-evident, that all men are created equal, that they are endowed with certain inalienable rights, that among these are life, liberty, and the pursuit of happiness, then it should also be self-evident that the abandonment of millions of people in modern societies to unemployment and destitution is a worse defilement of the earth than nuclear power stations. If the ethical force of the environmental movement can defeat the manufacturers of nuclear power stations, the same force should also be able to foster the growth of technology that supplies the needs of impoverished humans at a price they can afford. This is the great task for technology in the coming century. The free market will not by itself produce technology friendly to the poor. Only a technology positively guided by ethics can do it. The power of ethics must be exerted by the environmental movement and by concerned scientists, educators, and entrepreneurs working together. If we are wise, we shall also enlist in the common cause of social justice the enduring power of religion. Religion has in the past contributed mightily to many good causes, from the building of cathedrals and the education of children to the abolition of slavery. Religion will remain in the future a force equal in strength to science and equally committed to the long-range improvement of the human condition.

In the world of religion, over the centuries, there have been prophets of doom and prophets of hope, with hope in the end predominating. Science also gives warnings of doom and promises of hope, but the warnings and the promises of science cannot be separated. Every honest scientific prophet must mix the good news with the bad. Haldane was an honest prophet, showing us the evil done by science not as inescapable fate but as a challenge to be overcome. He wrote in *Daedalus* in 1923, "We are at present almost completely ignorant of biology, a fact which often escapes the notice of biologists, and renders them too presumptuous in their estimates of the present position of their science, too modest in their claims for its future." Biology has made amazing progress since 1923, but Haldane's statement is still true. We still know little about the biological processes that affect human beings most intimately— the development of speech and social skills in infants, the interplay between moods and emotions and learning and understanding in children and adults, the onset of aging and mental deterioration at the end of life. None of these processes will be understood within the next decade, but all of them might be understood within the next century. Understanding will then lead to new technologies that offer hope of preventing tragedies and ameliorating the human condition. Few people believe any longer in the romantic dream that human beings are perfectible. But most of us still believe that human beings are capable of improvement.

In public discussions of biotechnology today, the idea of improving the human race by artificial means is widely condemned. The idea is repugnant because it conjures up visions of Nazi doctors sterilizing Jews and killing defective children. There are many good reasons for condemning enforced sterilization and euthanasia. But the artificial improvement of human beings will come, one way or another, whether we like it or not, as soon as the progress of biological understanding makes it possible. When people are offered technical means to improve themselves and their children, no matter what they conceive improvement to mean, the offer will be accepted. Improvement may mean better health, longer life, a more cheerful disposition, a stronger heart, a smarter brain, the ability to earn more money as a rock star or baseball player or business executive. The technology of improvement may be hindered or delayed by regulation, but it cannot be permanently suppressed. Human improvement, like abortion today, will be officially disapproved, legally discouraged, or forbidden, but widely practiced. It will be seen by millions of citizens as a liberation from past constraints and injustices. Their freedom to choose cannot be permanently denied.

Two hundred years ago, William Blake engraved *The Gates of Paradise*, a little book of drawings and verses. One of the drawings, with the title "Aged Ignorance," shows an old man wearing professorial eyeglasses and holding a large pair of scissors. In front of him, a winged child is running naked in the light from a rising sun. The old man sits with his back to the sun. With a self-satisfied smile he opens his scissors and clips the child's wings. With the picture goes a little poem:

Aged Ignorance
Perceptive Organs closed their Objects close
Published 17 May 1793 by W Blake Lambeth

AGED IGNORANCE

*By William Blake, 1793. Reproduced by permission
of the Houghton Library, Harvard University.*

In Time's Ocean falling drown'd,
In Aged Ignorance profound,
Holy and cold, I clip'd the Wings
Of all Sublunary Things.

This picture is an image of the human condition in the era that is now beginning. The rising sun is biological science, throwing light of ever-increasing intensity onto the processes by which we live and feel and think.

The winged child is human life, becoming for the first time aware of itself and its potentialities in the light of science. The old man is our existing human society, shaped by ages of past ignorance. Our laws, our loyalties, our fears and hatreds, our economic and social injustices, all grew slowly and are deeply rooted in the past. Inevitably the advance of biological knowledge will bring clashes between old institutions and new desires for human self-improvement. Old institutions will clip the wings of new desires. Up to a point, caution is justified and social constraints are necessary. The new technologies will be dangerous as well as liberating. But in the long run, social constraints must bend to new realities. Humanity cannot live forever with clipped wings. The vision of self-improvement, which William Blake and Samuel Gompers in their different ways proclaimed, will not vanish from the earth.

June Goodfield, *An Imagined World: A Story of Scientific Discovery* (New York: Harper and Row, 1981).

INTRODUCTION

The first Auden quotation is from an essay, "The Fall of Rome," written by Auden in 1966 but unpublished until 1995. It was finally published with an introduction by Glen Bowersock in *Auden Studies*, 3 (New York: Oxford University Press, 1995), pp. 111–137. The passage quoted is on page 130. I am indebted to Professor Bowersock for bringing it to my attention.

The second Auden quote is taken from the chapter "Auden and the Liposome," in Gerald Weissmann, *The Woods Hole Cantata* (New York: Dodd, Mead and Company, 1985), p. 81. Weissmann gives the reference: W. H. Auden, in *The Place of Value in a World of Facts: 14th Nobel Symposium*, A. W. K. Tiselius and S. Nilsson (New York: John Wiley Interscience, 1970). Auden had a strong and well-informed interest, unusual for a poet, both in religion and in science. Weissmann has a strong and well-informed interest, unusual for a scientist, both in poetry and in art.

H. G. Wells, *The Time Machine*, ed. Frank D. McConnell (New York: Oxford University Press, 1977), p. 104. The novel was first published in 1895.

Note on the subsequent history of Onkel Bruno's oak tree. In 1996 the tree is still standing. He did not carry out his plan to replace it before he died, and it is now protected by state law as a historical monument. After the unification of Germany, his doctor son moved the family medical practice out of the village to a more prosperous nearby town. The son now lives in a modern house which he built in the garden beside the old house. The grandchildren are enjoying the oak tree, but not in the way Onkel Bruno intended.

chapter one

STORIES

James Leasor, *The Millionth Chance: The Story of the R101* (New York: Reynal and Co., 1957). The title refers to the statement made by Lord Thompson before he embarked on the fatal voyage of R101, "She is as safe as a house, except for the millionth chance."

Nevil Shute, *Slide Rule: The Autobiography of an Engineer* (London: Heinemann, 1954). The history of R100 and R101 occupies the middle third of the book, pp. 53–134. Shute's novel *No Highway* was published in 1948.

John McPhee, *The Curve of Binding Energy* (New York: Farrar, Straus and Giroux, 1974), describes the life and work of Ted Taylor.

For a more detailed account of the early days of Taylor's ice-pond projects, see F. J. Dyson, *Infinite in All Directions* (New York: Harper and Row, 1988), chap. 8, pp. 149–155.

chapter two

SCIENCE

My Oxford lecture on occultation astronomy was published in F. J. Dyson, *Quart. Journ. Roy. Astron. Soc.*, 33 (1992): 45–57. The lecture will be reprinted with an addendum to bring it up to date in a volume, *The Milne Lectures, 1977–1996* (London: Oxford University Press, 1996).

James E. Gunn and David H. Weinberg, "The Sloan Digital Sky Survey," Preprint IASSNS-AST 94/64, Institute for Advanced Study, Princeton, N.J. (1994). To appear in *Wide-Field Spectroscopy and the Distant Universe*, ed. S. J. Maddox and A. Aragón-Salamanca (Singapore: World Scientific).

Fritz Zwicky's Halley Lecture, "Morphological Astronomy," was published in *The Observatory*, 68 (1948): 121–143. See pp. 126–127.

chapter three

TECHNOLOGY

J. B. S. Haldane, *Daedalus, or Science and the Future* (London: Kegan Paul, 1923). The description of Daedalus is on pp. 46–48, the remark about Einstein on p. 11, the remark about applied science magnifying injustices on p. 85.

The experiment using a mouse gene to create a fruit-fly eye is reported by Georg Halder, Patrick Callaerts, and Walter Gehring, "Induction of Ectopic Eyes by Targeted Expression of the Eyeless Gene in Drosophila," *Science*, 267 (1995): 1788–1792. References are given there to earlier work by Gehring's group.

The theory that the Cambrian explosion of higher life forms became possible after the evolution of a two-stage pattern of morphological development is proposed by Eric H. Davidson, Kevin J. Peterson, and R. Andrew Cameron in "Origin of Bilaterian Body Plans: Evolution of Developmental Regulatory Mechanisms," *Science*, 270 (1995): 1319–1325.

For the evolution of eukaryotic cells, see Lynn Margulis,

Symbiosis in Cell Evolution (San Francisco: Freeman and Co., 1981).

For the two-stage theory of the origin of life, see F. J. Dyson, *Origins of Life* (Cambridge: Cambridge University Press, 1985).

Olaf Stapledon, *Last and First Men* (New York: Dover Publications, 1968), was originally published in 1931. *Sirius* (New York: Dover Publications, 1972), was originally published in 1944.

Aldous Huxley, *Brave New World* (London: Chatto and Windus, 1938).

Bruce Chatwin, *Utz* (London: Viking-Penguin Books, 1988).

Saul Bellow, Nobel Lecture, 1976. For this quote I am indebted to Clara Park, "No Time for Comedy," *Hudson Review*, 32 (1979): 191–200.

chapter four
EVOLUTION

Robert Forward, *Dragon's Egg* (New York: Ballantine Books, 1980).

The "cult of evanescence" is described by Stapledon in chapter 12 of *Last and First Men.*

The Carroll universe is described in J. M. Lévy-Leblond, "Une Nouvelle Limite Non-relativiste du Groupe de Poincaré," *Annales Inst. Henri Poincaré*, 3 (1965): 1.

My speculations about the possibility of life surviving forever in an expanding universe were published in F. J. Dyson, *Reviews of Modern Physics*, 51 (1979): 447–460. The case of a universe with critical density was considered by Steven Frautschi, *Science*, 217 (1982): 593–599.

For Gaia, see James Lovelock, *The Ages of Gaia: A Biography of Our Living Earth* (New York: Norton, 1988).

chapter five

ETHICS

The inscription on the Gompers monument was taken from a speech given at the International Labor Conference in Chicago in 1893. See "The Papers of Samuel Gompers, Vol. 3," ed. Stuart B. Kaufman et al. (Urbana: University of Illinois Press, 1989), pp. 388–396. For this reference I am indebted to Professor Joseph McElrath of Florida State University.

The Hardy quotation is on page 60 of *A Mathematician's Apology* (Cambridge: Cambridge University Press, 1940).

George F. Kennan, *The Nuclear Delusion* (New York: Pantheon Books, 1982), pp. 201–207.

The alert reader may observe that I omitted the words "by their Creator" from Jefferson's declaration. We may agree with the endowment of inalienable rights, even if we do not believe in a personal Creator.

Haldane's remark about our ignorance of biology is on page 50 of *Daedalus* (see note to Chapter 3).

William Blake's *The Gates of Paradise* was published in two versions, the first "For Children" in 1793, the second "For the Sexes" in 1810. The verses quoted here appeared only in the second version. See *The Portable Blake*, ed. Alfred Kazin (New York: Viking Press, 1946), pp. 273, 277, 697, 698.